RUNAWAY

To Jeff:

Thanks for reading the early draft. I hope you enjoy the final version.

A NOVEL

JOHN TOPPING

LONGSTREET PRESS
Atlanta

RUNAWAY

Published by
LONGSTREET PRESS, INC.
2140 Newmarket Parkway
Suite 122
Marietta, GA 30067

Printed in the United States of America

1st printing 2001

Library of Congress Catalog Card Number: 2001092693

ISBN: 1-56352-691-3

Jacket and book design by Burtch Bennett Hunter

To my wife, Laura, for her patience,
understanding and support.

✦

I would like to thank all of those who helped me through the
process of writing *Runaway*: Dawn, Julian, Craig, Tom,
Paige, JoAnne, Rob and many others, thank you.
Your encouragement and ideas
helped immensely.

RUNAWAY

CHAPTER 1

Thunder rolled across the Florida coastline as the EOS 5 satellite headed into orbit atop its Atlas booster. NASA designed the EOS 5, the fifth in the Earth Observing System series of satellites, to detect variations in the color and chemical composition of the top ten meters of the surface of the Earth's oceans. Upon obtaining its proper orbital inclination, EOS 5 deployed its instrument panels and began to transmit data to NASA's receiving stations across the globe. The receiving stations then relayed the information to NASA's data collection center in Arlington, Virginia.

❖

MCLEAN, VIRGINIA.

Jason gazed out the window of the South Wing conference room. He didn't notice the wind swirling the Autumn-hued leaves or hear it groan against the window. Instead his thoughts focused on their discovery and on its ramifications for the future.

He turned to face the members of his team. "So, everyone agrees. It's the runaway greenhouse effect."

Robert Wayneright looked up through the dark circles under his eyes. Jason wondered when Robert had last gotten a good night's sleep. "I can't believe we've done this! How could we let it happen?"

Jeff Nakamura, the doubting Thomas in the group, lowered his unlit pipe. "Robert, it hasn't happened yet. We haven't had time to test out the data. There were some anomalies..."

Robert shook his head. "The data's clear. Greenhouse gases and temperatures will rise gradually over the next fifteen years. After that, the increases will be dramatic. The oceans have stopped absorbing excess carbon dioxide. We've filled our planet's carbon sink and overwhelmed Nature's feedback mechanisms. There's nothing we can do."

Jason tightened his hold on the edge of the conference table. "I can't accept that. None of us can." Jason scanned the room. Jeff met his gaze. The other team members couldn't. Arthur Jacobson flipped through sheet after sheet of computer printouts. Robert Wayneright stared vacantly out the window. Natalie Ivanenko wiped mascara off her cheeks and looked down at her opened appointment book. He glanced at his watch. "I've got to meet up with Jack Watson in a few minutes. I'm not going to mention our discovery to him yet and expect all of you to keep this under wraps. Natalie, clear all of our appointments for the next 48 hours. Robert, I'd like a detailed summary of the data on my desk tomorrow morning. Address those anomalies that Jeff mentioned. Arthur, confirm that this isn't some kind of hardware glitch. I know, you've told me about the Omegatron before. Double check anyway. Jeff, you do the same with the software. We'll meet back here at 10:00am tomorrow."

Jason ran into Jack Watson in the corridor outside his office.

"Ready?"

Jason nodded. They walked out to Jack's car. "Quite a breeze, isn't it?"

"I hadn't noticed."

"You look a little tense. Nervous about meeting Congressman Barnes?"

Jason nodded again. As a senior member of the House Appropriations Committee and the impetus behind his project, Jason had a right to be nervous about meeting the Congressman. He only wished the meeting were the cause of his tension.

"Don't worry, Geoff and I go way back." Jack unlocked the door to his Cadillac. "One thing about Washington, Jason; it doesn't pay to drive a foreign car." Twenty-five minutes later they walked into the restaurant. The maitre d' recognized Jack and escorted them to their table. As they took their seats, the maitre d' rushed over to greet a tall, grey-haired man in a pin-striped suit. He pointed in their direction. After several stops along the way to talk with diners, the Congressman made it to their table. Jack and Jason rose to greet him.

Congressman Geoff Barnes was taller than Jason had expected. He dwarfed Jack and was several inches taller than Jason. Taking Jack's hand, the Congressman said, "Good to see you again, Jack. It's been what, two months since you shellacked me on the links?" Without pausing for a response, the Congressman turned and grasped Jason's hand "Who's this you brought along with you, Jack?"

Jason answered, "Jason Graham, Congressman. It's a pleasure to meet you."

"Graham, I've heard a great deal about you. In fact, you're one of the reasons EnviroCon got the contract." Jason glanced over at Jack. He hadn't heard that before. Now EnviroCon's offer made more sense. The Congressman paused to take a sip of his iced tea. "You've come aboard at a good time, Jason. Your NASA project is an important one. That data's been warehoused too long. It's time for us to put it to use." The Congressman paused and looked over Jason's shoulder. "Here comes my assistant now."

The woman who approached the table stood five feet seven inches tall with shoulder-length blond hair. Although she wore a conservative suit that disguised her figure, Jason knew the figure well. It hadn't changed.

"Sorry I'm late, Geoff. I had trouble contacting Congressman Gallager. Then, Congressman Friedman called. He wanted a status report on the EnviroCon project. Geoff, he knew about your appointment." Congressman Barnes rolled his eyes.

Jason and Jack stood to greet the assistant. "Jack, you've met Samantha Whitlock before, haven't you?"

"Of course I have. Nice to see you again, Samantha." Jack shook Samantha's hand.

Congressman Barnes began to introduce Samantha to Jason, but Sam stepped forward. "Hello, Jason. It's been a long time."

Jason, recovering from his initial shock, smiled in response, "Too long." Her eyes held the same sparkle he had noticed when they first met.

"Sit down. Please." Jason sat down slowly. Noticing Jack's inquisitive look, Sam briefly explained that she and Jason had been good friends at MIT. Jason didn't elaborate.

The lunch proceeded, but Jason had trouble focusing on the conversation. He found himself staring at Sam, frustrated because he couldn't talk to her. After picking at his blackened grouper, Jason excused himself to go to the rest room. On his way out, he almost bumped into Sam waiting just outside the door.

"How about a proper greeting?" Sam put her arms around Jason and pressed her lips against his. Jason responded passionately, but then broke off the kiss.

"Hold on. What are you up to?"

Sam smiled. "Not now. Let's get together soon. I'll explain." Sam handed Jason a business card. "My private line. Call me."

Sam turned and headed back for the table. After a moment's hesitation, Jason returned to the table as well. The Congressman had evidently taken as much time as he could out of his busy schedule. Jack was paying the bill when Jason returned.

◆

Jeff Nakamura left the converted mansion that served as the

EnviroCon offices a few minutes after Jason and Jack Watson. He, too, had an appointment with a Congressman for lunch. He traversed the DC suburbs, stopping at an elegant home in McLean, Virginia. A butler showed him to the study. Jeff pulled out his pipe and began to fill it as he sat down in a wingback chair. He pulled out his lighter and lit it before he remembered. He'd made that mistake once before, but the scent of cigar smoke had almost made him forget.

Jeff tapped his pipe in his hand as he waited. He had to get back to the EnviroCon offices, yet he knew that the Congressman would make him wait. When Jeff heard the click of heels on the hardwood floors, he rose to greet Congressman Joel Friedman. The Congressman took a seat opposite Jeff, pulled out a Churchill and lit it.

"What do you have for me?"

"They know."

"What do you mean?"

"They know. Jason Graham's software is too good. We met this morning. They know we're facing a runaway greenhouse effect." Jeff took a deep breath and waited. Congressman Friedman rose and turned towards the fireplace. He turned back quickly.

"God damn it Nakamura! What the hell have you been doing for the last 18 months. I installed you at EnviroCon to keep this from happening."

Jeff began to speak, but the Congressman's glare silenced him.

"I gave you access to the NASA data and NASA's state of the art Omegatron computer. I gave you access to my man inside NASA. I thought you could handle it."

Jeff tightened his grip on the arm of his chair. "Look, Congressman. I did my job. I falsified the data. Nobody else has figured it out and the press has been digging at the issue for years. But Jason Graham's program is too good. It rejected every false lead I planted. I did my job. You should have kept Graham off the team." Jeff met the Congressman's stare. The Congressman puffed at his cigar and then slowly withdrew it from his mouth. He smiled.

"You're right. Damn Geoff Barnes! He got my vote before bringing Graham into the mix." He took a couple puffs off his cigar. "Jeff, get back to your office, lay low, cover your tracks. I'll need a mole at EnviroCon. You're it. Now, I've got to think." The Congressman turned his back on Jeff as Jeff rose to leave.

CHAPTER 2

OCTOBER 7, 2006.
MEXICO CITY, MEXICO.

Meteorologists predict that Hurricane Erin will break the pattern of dry weather and stifling temperatures that have killed thousands in Mexico City and have caused massive crop failures. What people would usually view as a threat is now being greeted as a blessing. Although the one hundred-mile per hour winds and high waves will cause destruction along Mexico's coastal regions, the damage will be insignificant compared to the destruction already caused by the heat and the drought.

❖

MCLEAN, VIRGINIA.

Jason awoke with a start. He looked at his watch. 12:30 p.m. At least he'd managed to catch a few hours of sleep. The sofa in his office wasn't half-bad. He had almost gotten used to sleeping on it. He sat up, scratched his growing beard, and took a swallow from the Diet Coke. Flat. Jason frowned but took another swallow. He needed the caffeine.

Jason didn't run into anybody on his way down to the company health club. He jumped into the shower, letting the warm water play over the tense muscles of his neck and shoulders. Things were happening too quickly. The water pouring over him helped pull the stress out from his body, letting him focus on the situation. Jason knew, intellectually, that they were in real trouble. But deep down, in his gut, he could not accept his inability to do anything. He had hoped that either the data or the program was in error, but after all the checks they'd run, he knew they weren't. Jason refused to believe that a solution did not exist.

When Jason arrived at the conference room, all of the team members were present. They all looked haggard, Robert more so than the others. Jason knew that Jack would immediately sense that a problem existed. He'd have to be direct with Jack. He turned to Natalie and Robert, who were standing next to the main computer terminal, apparently checking the results of the latest model.

"What have we got? Any change?"

Robert responded. "No, unfortunately not. The new projections are worse. There's nothing we can do, Jason. In forty to fifty years, the world's population will drop to ten percent of its current level. Within three to four hundred years, the surface of the Earth will be totally uninhabitable. By that point I doubt that anybody will be around to observe the change. There's nothing we can do, Jason, nothing at all." Robert placed his face in his hands and held them there.

"But there's got to be something! I can't believe that the world is going to end no matter what. We just haven't had a chance to focus on a solution. We've been too focused on the data and the computer models."

Jeff Nakamura stood up and walked over to Jason. "I agree. I've focused all my energies on programming. I haven't had a chance to really think about what the data is telling us." Jeff picked up his unlit pipe. "My grandfather died at Nagasaki. I never knew him. But my grandmother, she suffered a lingering death at the hands of the cancers that grew within her. She used to tell me of the fire from the sky. For years I had nightmares about a nuclear war. They didn't go away until college. I wrote a thesis on thermonuclear war and came to the conclusion that all of the fears of nuclear winter were unwarranted. Although a large portion of the human race would have died from a thermonuclear conflict between the Soviet Union and the West, mankind would have survived. If mankind could survive a thermonuclear war, I can't believe that we will not be able to survive global warming. We've got decades to come up with a solution."

Robert turned towards Jeff. "We don't have decades." He punched a couple of keys. "Take a look at the screen. Even if the release of all greenhouse gases is brought to zero, starting today, the warming trend will continue. All that happens is that the temperature increases more slowly. And in the real world, what do you think the odds of cutting emissions to zero are? Even if we halt all man-made emissions, nature will continue to dump greenhouse gases into the atmosphere at a rate that will accelerate the pace of global warming. We're beyond the point of no return. Man has destroyed the global equilibrium and the planet is seeking a new one. There's nothing that we can do."

All of the heads in the room turned as the door clanged shut. Jack Watson walked into the room. "Why is everybody so somber? I'm only a few minutes early. And what do you mean, 'there is nothing that we can do'?"

Jason quickly walked over to meet Jack. "Jack, come on in. Have a seat."

Jack scanned the room. "I think I'd better stand. What the hell's going on? You told me nothing was wrong. Things look very wrong." Jason met his gaze.

"Jack, I told you the other day that nothing was wrong with the program, not that nothing was wrong. The programs we developed have worked perfectly. It's the data itself that presents the problem."

Jeff interrupted. "Cut the crap Jason. Jack, the Earth is in the early stages of a runaway greenhouse effect. The world is on the brink of collapse."

Jack looked at Jeff and then at the others in the room, stopping with Jason. "I hope that this is one of those practical jokes of yours, Jason, but if it is, it's not very funny."

"I wish it was, Jack. Jeff wasn't kidding."

"You mean that the world is coming to an end?"

Robert stood up. "This is for real, Jack. Look at me; look at all of us. We've been trying to punch a hole in our findings, but we can't."

Jack lowered himself slowly into a chair. "Start at the beginning."

When Jason finished, he turned toward Jack. "Well, Jack, what do you think?"

"I'm too numb to think." Jack looked around at the faces in the room. "We've all known that we've been causing grave harm to the environment for decades, but I never would have thought that we could do so much damage. Are you sure that both the program and the data are correct?" Before Jason could answer, Jeff stood up.

"Yes. You don't know how many times I checked Jason's work, both before and after we obtained the results."

Jack drummed his fingers on the table. "Then why hasn't anybody noticed this before?"

Waving the printouts in front of him, Arthur said: "People have predicted a runaway greenhouse effect since the middle of the twentieth century, but couldn't prove it. With this data they would have been believed. And back then there might have been something that we could have done. But the satellite data wasn't available at the time and until Jason's programs and the Omegatron, nothing existed to properly correlate the data."

Jason stood up and walked back over to the window. Looking at the leaves blown by the Autumn wind made the entire meeting seem surreal to him, yet he knew it was real. "I think that we should call a press conference."

Both Jack and Jeff began to talk at once. Jack glared at Jeff and Jeff conceded the floor to him. "You're premature, Jason. We need to confirm the results independently before we do that. What if you're wrong, the release of the data is liable to cause a panic, assuming it's believed."

"At least the data will be up for public scrutiny. Perhaps somebody will prove us wrong or will come up with a solution. Besides, if we're right, wouldn't you want to know?"

"Jason, I'll concede all of your points, but we need to go through channels. We can't drop this kind of bombshell on the public without adequate preparation. We should review the data with Geoff Barnes, the people at NASA and the people at the EPA. The indiscriminate release of the data would be irresponsible."

"Jack, I've been around long enough to know that once we begin to discuss things with the government, the information will leak out. Special interest groups will leak selected portions of the data, putting their own spins on it. The process will create more confusion than anything else. If we go public, we can control the process."

Natalie, who had remained silent to this point, interjected, "Jason's right. I know that you and Jeff disagree, but who are we to keep this information from the public?"

Jeff retorted. "You are dead wrong, Natalie. We must keep this close to the vest. Besides what difference will it make if we delay things a few months?"

While everybody else had been talking, Robert had been busily typing at the computer terminal. The lights dimmed and the screen along the wall lit up. Everybody turned towards the screen. Jeff piqued, "Who did that?"

Robert stood up. "I did. I can resolve this debate once and for all. Look at the screen. Chart 1, on the left, shows the death toll over the next thirty years under our baseline model. Chart 2, on the right, shows the death toll varying only one factor. As you can see, the death toll begins to climb in year two on Chart 2. By year fifteen, Chart 1 shows a five percent decline in population whereas Chart 2's decline is thirty-five percent. I'm sure all of you are curious about the one factor. Bear with me for a minute."

"When the data first came out I was depressed. I felt that we had an obligation to go public for all the reasons that Jason mentioned. Even if we are proven to be wrong at the cost of our reputations, at least I could sleep at night. However, when I came out of my initial fugue of depression, something about the data began to bother me. I didn't understand why the decrease in human population began to grow so dramatically after year ten. While the conditions began to deteriorate faster at that point, we have the technology to replant and transplant crops and should be able to handle the crop failures in the third world with the surpluses that the developed countries generally have. The inclement weather conditions and the reduction in global food production alone weren't enough to justify the size of the decline in human population. Then I realized that year ten was the same time that the changes in temperature and weather would become obvious. I asked the computer about the variables that led to the increased death rate in year ten. The answer startled me. The only change between Chart 1 and Chart 2 is the time that the public at large learns about the disaster. In Chart 1, the public learns about the disaster in year ten, as a result of the climatic changes becoming too obvious to ignore. In Chart 2, the public learns about the disaster this year, presumably by our going public."

Robert paused to look over at Jack and Jason. "As much as I would like to go public right now, we can't. If we do, we'll only accelerate the demise of civilization."

Jason couldn't think of anything to say; and, from the silence in the room, evidently nobody else could either. Keeping things quiet didn't lay right with Jason; yet what Robert said was logical. Why would people keep working if they knew it was to no avail? What incentive would they have? Once people began to believe the revelation of disaster, economies would grind to a halt. Some measures could be taken, but democracy could not survive. People would need to be forced to work. They would revolt and civilization would fail sooner. Jason knew that finding a solution, which he was convinced must exist, would take time. A public revelation of the disaster would end any real chance that he would have to find and implement a solution.

Jack interrupted Jason's thoughts. "You've all given me a great deal to think about. But I have to admit that what Robert said and what the computer confirmed, makes sense. We are going to need to keep this very quiet. One leak and we're dead. We'll be back to Chart 2. We need to bring our chief of security, Michael Barnwell, in on this. We'll need his input on how to keep this quiet and prevent leaks."

"You all look exhausted. Go get some rest. I'll brief Michael and then we can meet here Monday morning. However, do not discuss any of this with your loved ones. Make some excuse, any excuse, for your condition, but keep in mind that if this leaks out, you'll only be shortening the time that we have to respond to the crisis."

Everybody slowly got up from the table and left. Jason left through the break room to get a fresh Diet Coke. Jack followed right behind him. While Jason grabbed the Diet Coke, Jack fixed himself a cup of coffee. He stirred in a package of sweetener. "You know, for a minute there I actually blamed you. I was angry with you." Jason frowned. "I got over it pretty quickly, but if I could feel that way, how do you think people that don't know you might react? Bearers of bad tidings are rarely received well."

Jason turned abruptly, spilling his drink. "I don't give a damn about what people think! All I care about is finding a solution. I wish you and Jeff would stop thinking about covering your asses."

"Calm down Jason. You misunderstood me. I agree with you. This is much too important to let personal concerns cloud our judgment."

Jason relaxed his grip on his dented Coke can. "I'm sorry, Jack. I'm tired. Besides, if you're worried about my wanting to bring the data public, you don't have to be."

Jack put his arm around Jason's shoulder. "You and your team have shouldered quite a burden over the past few weeks. I would have blown up well before now. Go home. Try to get some rest. I'll get with Michael Barnwell."

Jason returned to his office to shut down his computer. He was glad he did when it reminded him of his next appointment; a date with Sam. She was meeting him at 7:00 at his apartment.

When he got home he went for a run to clear his head. Something kept nagging at him, but he couldn't put his finger on it. After showering, Jason dressed and popped open a bottle of Labbatt Blue. His doorbell rang at precisely 7:00. Sam had always been prompt.

Jason opened the door. Sam brushed past him and began to look around his apartment. Jason smiled remembering the day she walked into his apartment at MIT for the first time. "Can I get you a drink before we go?"

"Sure, I'll have a rum and ginger ale." Jason fixed Sam's drink and grabbed his beer. He handed Sam her drink and sat down next to her.

"I've picked out a new Indian restaurant in Georgetown. You do still like Indian, don't you?"

"Yeah. Sounds great. I haven't had any in ages." Jason paused as he took a swallow from his beer. "You really look great, Sam. You haven't changed a bit."

"Thanks. I appreciate that. You look like you've kept in pretty good shape yourself. Still playing tennis?"

"I did in California. But I haven't found anybody to play with here. Are you playing?" Jason's attempts to teach Sam how to play in college had resulted in one of their few fights.

"Believe it or not, I am. I've found a few people on the Hill that like to play. We should play together." A hint of a frown passed over Jason's face. Jason had never liked playing mixed doubles. "Don't worry. I bet that I'll surprise you." Sam smiled.

"Okay, let's play next weekend. We can play here or at my office. Believe it or not, they've got a court."

"Oh, I believe it. I've been there. It's quite an office."

"You're telling me." Sam finished her drink as Jason downed a last swallow of beer. "Ready?" Sam nodded. Jason picked up

Sam's jacket and met her at the door. Sam led Jason to a new English racing green Jaguar coupe. Jason hadn't seen one on the road yet, but wasn't surprised to find Sam driving one. Jason turned to Sam while putting on his seat belt. "Aren't you breaking an unspoken rule in Washington about foreign cars?"

"Not totally, I'm just stretching it. Since Ford owns Jaguar, I consider it an American car. Besides, I'm not a politician. Sure, if all of Congressman Barnes' staffers drove one, he might catch some flack. But I'm probably the only one that has a wealthy father that likes to give his only daughter expensive cars."

"How's your father? I haven't seen him since Thanksgiving the year before I went to Cal Tech."

"He's doing well. He supposedly retired last year, but you know how he is. He's down to a 60-hour week. If he lived more than an hour out of town, I don't know when we'd get to see each other."

"An hour the way you drive or an hour the way most people drive?" Sam threw Jason one of her knowing smiles and accelerated around a curve, pressing Jason against his door. Jason decided that the drive would take him at least 75 minutes.

When they arrived in Georgetown, Sam pulled into a small parking lot on P Street, just off Wisconsin Avenue and right behind the restaurant, a feat that Jason only later learned to appreciate. While the decor left something to be desired, the restaurant was comfortable and relatively quiet for a Saturday night.

"So what've you been up to since leaving MIT?"

Sam put her drink down. "Not much. I tried a few things. I was about to take my father up on joining him in business when I met Geoff Barnes. He was running his reelection bid. He impressed me, so I volunteered to work for his campaign. He won. One thing led to another and here I am, on Geoff's staff."

Jason smiled. "You're not only on his staff, you're his chief of staff."

"What can I say, Geoff's a good judge of talent."

"How'd you find out that I was coming to Washington?"

"Just got lucky. Geoff's Committee oversees the NASA budget.

EnviroCon made a pitch for the business. Seems they had some brilliant grad student from Cal Tech ready to run the program. That peaked my interest. From there it didn't take much for me to find out the details. Your arrival. Where you'd be living. Simple."

"Well, I'm glad that I piqued your interest after all these years. I was sure that you'd be married by now."

Sam took a sip from her drink and smiled. "No, I had one relationship, but it didn't work out. And to answer your next question, no, I'm not dating anybody right now. What about you? Did you meet any intriguing blondes on the West Coast?" Sam leaned forward and gazed into Jason's eyes.

"Not really." Jason looked down at his plate and shifted the food around. "You know, this tandori is great. Would you like some?" Jason moved his plate toward Sam.

"No thanks. How 'bout you? Want one of these onion bhajis? I noticed you looking over here enviously."

Jason chuckled, "That wasn't what I was looking at. But, sure, I'd love one."

"So, how are things going on your project?"

Jason hoped that the long nights of poker at MIT had trained him to keep a straight face. "Good, but I can't go into detail."

Sam nodded. "I know, security. But do you like it?"

"It's not what I expected, but yeah, I'm glad I took the job."

Sam paused. Jason was worried that he'd given something away. Sam knew him so well. "Good. Now I know why you're never home. One thing, though, I hope that you'll have time for me."

Jason was wiping his chin when Sam's words sunk in. He dropped his napkin, leaned over to get it and banged his head on the table.

"Are you all right, Jason?"

"I'm fine. You know, I never really understood why we broke up. In fact, I haven't been able to get you out of my head." Jason extended his hand toward Sam across the table. She took it. Jason cleared his throat. "How about coffee and dessert?"

"Sure. But let's go somewhere else. There are several good places along Wisconsin Avenue."

They walked down the street arm in arm. A sudden gust of warm, damp wind hit them. They ducked into a small café and grabbed a table. Jason ordered the cheesecake, a cup of expresso for Sam and a keemun tea with a snifter of amaretto for himself. Before the waitress returned with their order Jason felt a tap at his shoulder. He turned his head.

"Jason, it is you. I couldn't tell from across the room, but I thought it was you."

"Nice to see you, Jack."

Jack Watson and an attractive brunette were standing behind Jason. "I don't believe that you've met my wife. Michelle, this is Jason Graham."

Jason stood up and shook Michelle's hand. "A pleasure to meet you, Michelle."

"My pleasure, Jason. Jack has told me a great deal about you."

Sam kicked Jason's ankle under the table. "By the way, this is Samantha Whitlock." Sam and Michelle shook hands and began to chat. Jack gave Jason an inquisitive look and leaned over. "I'm glad to see you, but surprised. This place is off the beaten path. Besides, I thought I told you to get some rest."

Jason grinned. "Some things are more important than rest. And we hadn't planned to come in here. We ducked in to get out of the weather."

"What do you mean?" Jack looked puzzled.

"A gust of wind caught us. It felt as if it was about to pour. See." Jason glanced over his shoulder. Rain pelted the window of the cafe.

"But it wasn't supposed to rain." Jack shook his head. "Strange weather. Must be El Niño." Jack laughed. "Sorry, I forgot that you wrote the book on El Niño."

Jason lost track of what Jack was saying as he felt Sam's hand slide onto his leg under the table. He began thinking up excuses to tell Jack why they had to leave.

CHAPTER 3

OCTOBER 23, 2006.
CORAL GABLES, FLORIDA.

A weather satellite orbiting over the Atlantic detected a rapidly organizing tropical wave off the west coast of Africa. It transmitted the data directly to the National Hurricane Center in Coral Gables, Florida, where it raised the eyebrows of Emanuel Ortega, a young meteorologist on duty that morning. The hurricane season had already spawned half a dozen hurricanes, including one category 5 that had caused extensive damage in the Yucatan Peninsula. Despite the lateness of the season, water temperatures and other conditions remained favorable for the development of additional strong storms. Ortega brought the information to the attention of the supervisor, who glanced at the data briefly and told him to file it away and watch for further developments.

✦

MCLEAN, VIRGINIA.

Jason awoke early feeling great. After renewing his relationship with Sam, he felt as if a burden had been lifted from his shoulders. Sam gave him confidence and support. In fact, he was sure that his newfound confidence is what had enabled him to come up with the plan. He had worked hard all weekend, trying to come up with something prior to the Monday team meeting. While jogging early Sunday morning he finally realized what had been nagging the back of his brain. He spent the rest of the afternoon and evening on line validating his idea.

When Jason arrived at the office, he went to the cafeteria to grab a sausage biscuit and a Diet Coke. He bumped into Jack Watson in line. "Good morning Jack." Jack turned his head. He was now the unshaven one with bags under his eyes. "You look tired. Get any sleep over the weekend?"

Jack grinned wryly and shook his head. "You know the answer to that. I kept thinking about the beach. I'm really going to miss going to the beach. Michelle and I first met on the beach in Boca Raton during Spring Break. I proposed to Michelle five years later on a beach at Paradise Island in the Bahamas. We have lots of things to worry about, but the image of the destruction of all of those beaches got to me. I got out of bed last night, spent a few minutes staring at the kids sleeping and then went out to our porch to watch the stars and think. Michelle found me there in the morning. We sat together and watched the sun come up."

Jack looked over at Jason, who no longer had bags under his eyes and was clean-shaven. "It looks as if you didn't have much trouble sleeping last night, Jason. What's your secret?"

"Just lucky I guess. Besides, I was so exhausted after the weekend that I doubt anything could have kept me awake. How are things going with Michael Barnwell?"

"Good. He took the news in stride. But then, he's always been difficult to read."

When they arrived at the conference room, Arthur was already there. Jeff and Robert walked in a couple of minutes later. Michael and Natalie entered together at precisely 9:00.

Jack started, "If everybody will take their seats, we'll begin." Jack took a last sip of his coffee, leaned forward in his chair and cleared his throat. "We all know why we are here and the gravity of the situation that we face. What I would like to accomplish in this meeting is to come up with a game plan. I want all of you to speak freely and raise whatever points you see fit. But let me remind you, this is not a democracy. I will make the final decision. Period."

Jack looked down the table at Robert Wayneright, "Robert, you are the closest thing to a climatologist here. Why don't you start?"

Robert put the pencil down that he had been idly tapping against the table. "Our planet has departed from the climatic pattern that it has followed for the past several million years. We are, or I should say were, in the middle of an interglacial period. Studies of ice cores and ocean sediments indicate that the planet has experienced cycles of glaciation and warmth for several hundred thousand years. A number of feedback mechanisms have kept the planet's climate in relatively stable cycles. Unfortunately, the effects of man and industrialization over the past two centuries have upset the balance. We are no longer at equilibrium. The increase in carbon dioxide and other greenhouse gases has tipped the balance toward a runaway greenhouse effect."

Robert paused for a moment to take a swallow from his coffee. "Whether the planet will reach a new equilibrium before the Earth turns into another Venus is a question I can't answer." He paused to look down at his notes.

"We face an immediate crisis. Within ten years, twenty at the outside, conditions on the surface of the planet will begin to rapidly deteriorate. Droughts and killer storms, the likes of which we have yet to face, will become the norm. More than one-third of the population of the planet is located in coastal regions that will suffer the dual impact of rising sea levels and severe storms. Finally, as a result of the increase in heat combined with an increase in the amount of carbon dioxide and other pollutants in the atmosphere, breathing will become difficult over vast areas of the globe. Imagine being trapped in the Los Angeles Basin after a high-pressure system has held the air in place for several weeks during the height of the summer. Conditions in many locations will be far worse."

"Jack, you asked whether any of us have any insights. I wish I did. I have racked my brains, but don't even know where to start. We are in deep shit!"

Ignoring the last comment, Jack cleared his throat. "Thanks Robert. Before we go on, I'd like to address our next step. Despite the models, we need to bring a few people in on our secret. We need to go to the top with this, but we've got to do it right. First, I'll approach Congressman Barnes. He's loyal, he knows us and he knows how to keep things quiet. He'll be a strong ally for the future."

"Once we convince Congressman Barnes, we'll need to approach someone in the administration. Gene Samuelson is my choice. As National Science Advisor, he is in the right position to gain the ear of President Wright - assuming we need him. Besides, he was a classmate of mine. Other than that, we'll play it by ear. The fewer people that know the whole truth, the better. Washington is a town known for leaks. This is one leak nobody can afford."

Natalie interrupted. "What about the President? Shouldn't we go to him immediately."

Michael Barnwell chuckled at the suggestion. "No Natalie, that would be a mistake. The President is a bad risk. He has a habit of leaking information that's not in the public's best interest to know. The trade negotiations with China were a done deal until that last press conference of his. Besides, we don't need him."

"The democratic process is anything but perfect. There are some things that a government must do in secret. You've all heard of the "black" segment of the Pentagon's budget. Well, buried within that budget are special projects known as "double-black" projects. Even Presidents don't know about them. I've worked on double-black projects in the past. If ever a project existed that warranted a double-black security designation, this is it. The President does not have a need to know."

Michael's revelation shocked Jason. He knew about the "black" section of the Pentagon's budget. But the President and several members of the Congress were kept briefed on those "black" projects. The hypersonic stealth replacement for the SR 71 Blackbird had been developed as a black program. Jason had also read about several acts undertaken by the British during World War II that they had finally declassified. But Jason had never heard of a double-black project. Jason deplored the thought of unchecked power in the hands of a very few people. The risk of abuse was too great. Despite his intellectual agreement with the need to keep their discovery secret, he still had a nagging fear about it.

Evidently several other members of the team were shocked by Michael's revelations as well. The sound of Robert Wayneright's raised voice aroused Jason from his thoughts. "I don't care, damn it! It just isn't right. I know what the data says. But to keep something like this from the President! Who are we to sit in judgment over the Country and the entire world?" Fortunately, Jack intervened. He walked over to Robert and whispered a few things in his ear that managed to settle him down and then went back over to his seat. "I know how all of you must feel. The strain of keeping this information wrapped up is difficult, but we don't have a choice. As to Michael's revelations about so-called double-black projects, I don't care what was done in the past. We must focus on the future. As Michael observed, if ever there was a need for a double-black project, this is it. We can not afford a leak ... and I hope that all of you understand that. For now, let's focus our energies on finding a solution to this mess, not arguing amongst ourselves." Jack paused for several seconds. "Now, do any of you have anything positive or constructive to share with us before we go on?"

Jason decided the time was right to preview his idea. "Jack, I believe that I may have some good news." All of the eyes in the conference room turned towards Jason. "I've done a lot of soul searching over the past couple of weeks. I could not accept our impotence. I thought through the ramifications of our discovery and at first tried to think of what we could do to stop the climatic changes from occurring. But I kept going back to the results of our models. Even the impossible, zero growth, buys us but a few more decades."

"Then, in my frustration, I began to think about what we could do. I realized we couldn't return the Earth back to the way it used to be. We just don't have the time or the technology to do it. I briefly toyed with the idea of terraforming Mars, but it's not practical given our time frame. I then started to think about setting up colonies on Mars and moving people there so that at least some people would survive. While the idea has some merit, we don't have the time or the money to move enough people to Mars to make a difference."

"Finally, it dawned on me. I remembered a book I'd read in high school called the High Frontier. It proposed building space colonies at Earth's Lagrange points, which, for those of you that don't know, are the only truly stable orbits in the Earth-Moon system. I almost dismissed the concept, but just couldn't. The book had been very persuasive. I went online and pulled up a few articles on the subject. Building space colonies at Earth's Lagrange points is very doable. Even the timing works. From the figures I ran through, we can construct an initial 10,000 person colony within five years. Once the first colony is operational, growth will progress rapidly. We can build larger colonies and increase the population to several million ten years from today."

Robert began busily scribbling notes on a pad. Jeff sat reclined in his chair with his hands behind his head. Jason could tell that both Arthur and Natalie were on the verge of interrupting him, while Jack's eyes never wandered from him. "I'll give all of you a chance to poke holes in it later, but for now, give me a chance to finish."

Jason pulled several sheets of paper out of his jacket pocket. "I prepared a brief outline last night." Natalie started to get up.

"Don't get up, Natalie. One thing I learned from the courses that I taught as a graduate student, never hand out anything while talking. People read what's in front of them and don't listen."

"We need to take a long — hard look at my proposal, but I think that it will work. Granted, only an insignificant percentage of the globe's total population will be able to live on the Colonies, but civilization itself will not be lost. I have confidence that if we succeed in creating a self-sufficient civilization on L5 colonies, ultimately, we will be able to reclaim the planet. Perhaps in time to save a good chunk of the population."

Jason leaned back in his chair, the squeak breaking the silence. In looking around the room, Jason knew that he had already won several converts. Robert was one. He busily scribbled notes on the pad in front of him. If nothing else, Jason had managed to pull Robert out of the depths of his depression. All Robert needed was some glimmer of hope. Natalie, eyes bright and focused on Jason, was another.

Jeff Nakamura, as usual, looked like the doubting Thomas of the group. Jason believed that while Jeff would present problems, he would come around to Jason's position. Jason couldn't read Jack's reaction at all. Jack had the uncanny ability to suppress surprise. It was a talent that lent itself to Washington. Despite Jason's ability to remember every card played in a deck, he doubted that he could beat Jack in a game of poker. And Michael Barnwell, he'd probably beat Jack. Jack allowed everyone a few minutes of reflection before breaking the silence.

"Jason, while I appreciate your genius with a computer, you didn't focus on the practical side of our problem. Assuming for the moment that we are technologically able to construct enough colonies to maintain a civilization as things degenerate on the planet, how the hell are we going to get people to support the concept? How will we get it built? I have worked and lived within the Beltway long enough to know how things get done. You are talking about a massive commitment. How will we ever get it funded? Even assuming that we could make the disaster we face public, the public would still not get behind the project. After all, as you said, we are talking about saving a few million people, max. Who will choose those that are to survive and those that are not? Everyone will fight for a spot. Your idea may have technical merit, but it will not work."

"Jack, I'm not that naive. I know we're facing an almost insurmountable task, but I have confidence in the team. You've previously dealt with and solved political problems that initially appeared impossible. Given what's at stake, I'm sure you can cut through the political BS for us. Besides, what choice do we have?" Jason paused and took a sip of his tepid Diet Coke. "When I was a kid I read alot. I loved mysteries. We are not attempting to solve a mystery, but we are facing a seemingly impossible task. As Sherlock Holmes used to say: 'once you have eliminated the impossible, whatever remains, however improbable, must be the answer'. What I propose is improbable — not impossible. For now, it's all we've got. I'm counting on all of you help get the colonies funded and built. If any of you have a better solution, I'm all ears."

Jason leaned back in his chair. He knew that he'd have trouble convincing them of the merit of his idea, but what he had said was true, they had nothing else to go on. Besides, he truly believed it could be done. Jack stood and faced Michael Barnwell. "Michael, how much money can we spend covertly?"

"I'm not certain, but my guess is that the government hides ten to twenty billion dollars a year in double-black projects and another fifty to seventy billion in a black project." Jason let out an audible sigh. The number was high enough. Michael's eyes quickly focused on Jason's. "Don't sound too relieved, Jason. That's the total out of a four trillion dollar budget. Most of it's already allocated to other projects and is spread between the various military services, the President's office and the CIA. It will be very difficult to supplant those projects."

Jack stood up, his hands in his pockets. Jason watched as he paced back and forth. Jason knew that Jack's support was crucial. Jack's eyes focused on Jason. "Jason, how much do you estimate it will take to build these colonies?" He had him.

"Roughly 80 to 100 billion. Given maximum funding and a crash program, we could finish the work in just under four years."

Robert broke into the conversation. "Jack, Jason's overstating the cost of the project. Given our current launch capabilities and the techniques for space-based construction developed by the Russians, the cost of constructing the first habitat should be twenty to forty percent below Jason's projections. We've got a lot of work to do analyzing Jason's suggestion, but I think he's onto something."

Jack walked over to the window. The room remained silent as he stood there, gazing out. Jason scanned the room. All of the eyes in the room other than Michael Barnwell's were on Jack. Michael, like Jason, was looking from person to person, assessing their reactions. Michael's eyes caught Jason's for a minute and then moved on.

Jack turned back to face the members of the team. "Let's move forward with Jason's idea. I'd like each of you to think about the proposal and to come up with a list of the ten most significant difficulties and the ten most significant benefits of Jason's proposal. I'd also like all of you to propose solutions to each of the difficulties that you list. Please keep your assessments short and do not consult with each other. You'll have plenty of time to work together later. For now I'd like each of you to work independently. And I'd like to see your reports by 1:00 p.m. tomorrow. If any of you have any other proposals, I'd like to hear them as well. I hate to have all of our eggs in one basket on this one. By the way, I've taken the liberty of naming this project, "Project Runaway". Now, unless anybody has anything to add, I'll see all of you here tomorrow at 1:00 sharp."

❖

Jason returned directly to his office and flipped on his terminal. He had one piece of E-mail. "Jason, we need to talk. I'll be back in town tonight. Call me. Love, Sam." Jason read the message one more time and then erased it. He really missed Sam and relied on her support. "Shit!" Jason remembered the other thing he'd forgotten. The party at Sam's father's house. How could he have been so stupid! Jason called Sam's office. Sam's secretary, Dawn, told him that she'd left town with the Congressman, but that she was due back today. No, she didn't know the flight number. Jason hoped he hadn't blown his relationship with Sam. He'd promised her that he'd make time for her and yet, he'd forgotten that damn party. Jason accessed the airline reservation system computer but still didn't come up with a flight. He accessed Congressman Barnes's computer. It took Jason a few minutes to hack the Congressional security system. Once in, Jason located Sam's itinerary and retrieved her flight number.

He then hacked into the Washington National Airport computer and programmed a message into it for Sam. He enjoyed beating the Congressional and airport computer security systems. He hadn't hacked a computer in some time and was glad he hadn't lost his touch. Jason smiled as he imagined Sam's reaction to the message and their subsequent meeting. He hoped it would be enough.

Jason completed his report at 4:45 and left the office. On the way to his car, he noticed two people leaving in a small blue sports car. It looked like Michael Barnwell and Natalie, but he couldn't be certain. After arriving at his apartment, he took a quick shower, left a note on the door just in case he missed Sam at the airport and left for the airport.

WASHINGTON, D.C.

The private line rang in Congressman Joel Friedman's office. His secretary reached to pick up the line but he dismissed her. "Yes?"

"Congressman Friedman, it's Jeff Nakamura."

"Nakamura. What the hell are you doing calling me here?"

"No choice. It was now or never. The good news is they've decided to keep this quiet. Graham's program predicts that the release of the news will only accelerate the crisis. I didn't even have to argue the point."

"That is good news. I was concerned about our ability to keep this quiet. How are they going to contain the information?"

"That's one of the problems and the reason I called now. Jack Watson brought Michael Barnwell into the loop. He'll shut the lid tight on this. I won't be free to call you."

"Not a problem. Keep your head down. I'll make contact with you if I need you."

Congressman Friedman hung up and began to dial another number. A woman answered the phone, "General Maxwell's office."

"Get the General."

"Sir, I'll need to tell him who's calling."

"Tell him, Augustus." Congressman Friedman didn't have long to wait.

"What is it Joel?"

"Cancel operation silencer."

"You sure?"

"Yes, they're cooperating."

"Good. I was concerned about covering something like this up."

"I agree. They're bringing Geoff Barnes in. He's sure to bring you in given your position in Project Starcom. I'll enter the picture later, when Geoff needs my help." Joel Friedman hung up, picked up his cigar and puffed away.

On the flight back from New York, Sam fidgeted in her first class seat. She was thinking about the messages that she had left Jason, first at his apartment and then at his office. While she was still angry with him for not showing up at her father's, she regretted not having confronted Jason face to face. But what really bothered her was that her father had taken Jason's side. In a way Sam had reacted as much to her father's defense of Jason as to Jason's cancellation. If her mother had been alive, maybe she would have supported Sam.

The stewardess interrupted Sam's line of thought. She asked if either she or Geoff would like another cocktail. The plane was fairly close to final approach and this would be the last call. Geoff declined, having just finished a bourbon and water. Sam asked for a vodka and grapefruit, her second.

Geoff turned to Sam. "Are you all right? You've been rather tight-lipped this trip."

"I'm fine, Geoff, really. You know how I hate these fund raisers. Why do you bring me along? Virginia's far better at them. The press is right, she's your greatest asset."

"And you're a close second. Besides, you're my good luck charm. Ever since you joined my team, I haven't lost a battle. You keep me out of trouble with the press and know how to handle yourself among the elite. I'm the son of a poor teacher. I'm never comfortable at these $1,000 a plate dinners."

Sam laughed. "Don't give me that. You probably went to more parties than I did growing up." Geoff gave Sam one of the disarming smiles that had become one of his trademarks.

"I got you to laugh, didn't I. Besides, while I promised I wouldn't tell, Jimmy has a crush on you. He wouldn't forgive me if I came home without you."

It was Sam's turn to smile. She always stayed at the Barnes' home when the need came for them to travel through Princeton. The tradition had started due to budgetary constraints on the campaign committee. Sam had offered to pay for a room at a local hotel from her own pocket, but Geoff wouldn't hear of it. She had first met Jimmy when he was fifteen and on spring break from Lawrenceville. She remembered him as a wiry and energetic, but shy youth. Now she understood his shyness.

"It's too bad that Virginia had to stay in Princeton."

"I know, but with Thanksgiving coming up and with her mother's illness, it's best that she stay in Princeton. Besides, it gives me a great excuse to get out of D.C. as frequently as possible."

The stewardess came by and reminded them to fasten their seat belts. The plane was on its final approach into Washington National. Sam's thoughts returned to Jason. She had fallen for him during her first year at MIT. At the time, Jason had been dating Jennifer Chang, a knockout on the tennis team. Sam finally got Jason's attention by edging him out of the top slot in Professor Haven's environmental engineering class. After that, it was easy. Sam had planned to marry Jason, but his workaholic nature kept getting in the way of their relationship. After his move to California for graduate school, Sam gave up. Now Jason was back and Sam thought that she had learned to accept Jason's dedication to work. She had her own career and she loved it. Before renewing her relationship with Jason, it had absorbed all of her attention. Yet his missing the party at her father's really annoyed her. She had left several messages in anger. Now she regretted it.

After deplaning, they went to baggage claim. A page startled Sam: "Will Miss Samantha Whitlock please report to the information desk." In all her travels, she'd never been paged in an airport before. Geoff heard the page and offered to go with Sam, but she declined and asked if he could retrieve her bags and wait for her until she returned. Geoff agreed.

Sam walked briskly up the stairs back into the ticketing area. When she got to the information desk, she found a bored-looking man in an airline uniform sitting behind the desk. Next to him sat an enormous stuffed panda bear with a large red bow tied to its neck. "Excuse me, My name is Samantha Whitlock, I was just paged."

"Whitlock? That's right, this is for you." With that the man turned, wrapped his arms around the panda bear, lifted it over the counter and handed it to Sam. With difficulty, Sam wrapped her own arms around it and placed it on the floor. She guessed it was from Jason but didn't find a note. "Was there a note to go along with this?"

"No, ma'am. It arrived about 15 minutes ago. The delivery-man said that he was told to bring it here. Because there was no addressee, we checked it pretty thoroughly."

"Then how did you know to page me?"

"I didn't page you, ma'am. Somebody else must have. But after you were paged, I got a message on my terminal telling me to deliver the bear to a Ms. Samantha Whitlock. You're here and here's the bear." A skycap arrived with a luggage cart. "I got a message to meet somebody and then to go down to the baggage claim area."

The man behind the counter shook his head. "I didn't call for you, but you can give this lady a hand. By the way, who told you to come over here?"

The man took off his hat and scratched his short-cropped black hair. "It was a bit strange, actually. The screen behind our counter beeped a few times. I looked down and the message appeared on the screen, replacing the usual flight schedules. I thought that you might be able to tell me how you did that."

The man behind the counter shook his head again. "Beats me." Sam had a broad grin on her face. She gave the bear a hug and then asked the skycap to carry it for her down to the baggage area. When she got there, Geoff was gone, but a note flashed briefly across the screen located over the baggage carousel: "Sam, meet me outside."

If the skycap had noticed the message, he didn't say anything. Sam told him to take the bear outside. When she got out there, Jason was standing next to a limousine with the door opened, holding a glass of champagne in one hand. He winked at her. "Surprised?"

Sam ran over to Jason and threw her arms around him. "Oh, Jason, I'm so happy to see you." Jason pushed her gently away from him, put the glass of champagne down on the roof of the limo, and placed his hands gently on each of her cheeks: "I love you, Sam." He then proceeded to give her a passionate kiss. Jason handed the glass to Sam and helped her into the limo. He had the skycap load the bear into the back of the limo, gave him a generous tip and climbed into the limo himself. Sam snuggled up next to him.

They kissed for a minute and then Sam pulled away. "I'm really sorry, Jason. I overreacted."

"Don't say it, Sam. I ignored you. That's something that I promised that I wouldn't do anymore. But, something came up at the office and I just forgot about everything else. I love you desperately and am sorry if I hurt you."

"I know, Jason. I know how absorbed you get in your work. It's one of the things that I love about you. I'm sorry, but I was so looking forward to spending the weekend with you at Daddy's. When you didn't call or return my calls, I let my imagination run away with me. Daddy taking your side didn't help any either. By the way, what were you so busily working on at the office?"

A cloud briefly flitted across Jason's face. "I'll tell you later. For now, let's forget about work."

Sam jumped. "Work! I left Geoff at baggage claim, waiting for me!"

"Don't worry. I met Geoff and told him that I'd take care of your baggage and make sure you got home. But tonight is our night." Jason leaned over and kissed Sam.

The limo dropped them off for dinner at Otto's, a small Northern Italian restaurant in McLean. They split an appetizer and their main course. When the Secretary of State walked in for dinner, they were the only patrons that didn't notice him.

They had finished their first bottle of Chianti and were well into their second when they finished their main course. Sam excused herself. When She returned, Jason stood up to let her into the booth. He seemed nervous. He took a large swallow of his wine and then took her hands in his. "Sam, I've thought about this a lot over the past few days. I don't want to lose you. Sam, will you marry me?" The hint of a tear appeared in the corner of Sam's eye. She looked in Jason's eyes and nodded her head. "Yes, Jason, yes. I will. I've waited years for you to ask."

They left the restaurant and got back in the limousine. Because of their passionate kissing, Sam did not notice where they were going. When she finally did take note, they were well into the Virginia countryside. "Where are you taking me?"

"A little bed and breakfast that you'd mentioned to me several times." Sam snuggled up to Jason for the remainder of their trip out to Middleburg.

CHAPTER 4

After several years of negotiations, the government of Indonesia ratified the Rain Forest Preservation Treaty of 2000. Global pressure resulting from boycotts of the sale of Indonesian lumber had finally succeeded in convincing the last major holdout government to execute the Treaty. While logging is not absolutely prohibited, the Treaty governs the harvesting of tropical woods and prohibits clear cutting. The Treaty requires that each signatory government impose a tax on exports of its lumber and that they utilize the proceeds to reforest areas previously devastated by clear cutting.

Despite attempts on the part of member nations, loggers continue to clear-cut forests in violation of the treaty. Brazil's government has imposed fines on cutting timber in violation of governmental permits; however, very few inspectors are available and the Amazon forests have continued to decline since Brazil ratified the Treaty in 1999. Although the rate of decline has diminished, virtually one-quarter of the forest is now a useless wasteland. Despite their appearance, most tropical rain forests sit on poor quality land. The tropical rain forest ecology is extremely efficient and utilizes virtually all of the resources at hand. Once a rain forest is stripped, however, the land will not support cultivation for more than a few years.

✦

MIDDLEBURG, VIRGINIA.

As usual, Jason awoke early. Trying not to wake Sam, he climbed out of bed. His head hurt a bit so he popped a couple of aspirin and got in the shower. He had ordered breakfast in their room for 8:00. As he finished shaving, he heard a noise and looked over to see Sam slipping out of her robe. Jason's eyes wandered appreciatively over his future wife's body.

"Care if I join you?" Jason only nodded, but his body responded. "I see that you do." Sam stepped into the shower. He greeted her with a hug as she reached down and rubbed his erection. Jason lifted her up, onto him.

After getting out of the shower, Jason heard a knock at the door. He pulled on a bathrobe and went to open the door. A waiter brought the tray of food in and set up the table. Jason tipped the waiter, poured Sam a cup of black coffee and poured himself a Diet Coke. He placed the single red rose in Sam's chair and then lifted the metal heat covers to the dishes and looked over the breakfast. They had prepared scrambled eggs, cheese grits, biscuits with sausage gravy, two half grapefruits, toast and jelly. Before replacing the metal cover, Jason inhaled deeply, relishing the odor.

Sam entered the room wrapped in a bathrobe, drying her hair with a towel. She walked over to the table. Jason handed the rose to her, took her in his arms and kissed her. "I love you, Sam." Sam grinned broadly as she sat down at the table. "I'm in shock. Did you really ask me to marry you last night?"

"Yes, Sam, I did; and I'll do it again. Will you marry me?" Sam stared into Jason's deep brown eyes. "Yes, Jason, you know I will. I've wanted to and have imagined marrying you ever since our first date at MIT." She leaned over and kissed him. "We need to decide about when we want to get married."

"Pick a date. I'll get married whenever and wherever you want."

"If you don't mind, I'd like to talk it over with my father. Why don't we ride out to his house today and break the news to him?"

"That's a great idea, and I'd love to. I just can't. There's something else we need to discuss. But I need you to promise me that you won't repeat what I'm going to tell you to anybody, not your father or Geoff." Sam put down a fork full of eggs. Her face, which moments before had appeared so happy, now bore a small frown.

Jason took her hands. "Do you promise?" Sam nodded. "Don't look glum. This has nothing to do with our wedding or our relationship." Jason did not vocalize his thought about how seriously it did affect their future. "I'd like to tell you why I didn't make it to your father's party. As you suspected, it involves work." Jason took a minute to take one last bite of his biscuit and a sip of Diet Coke and then described what he and his team had discovered.

Sam recovered from the news quicker than Jason would have thought. "Hold on, Jason. You lost me there. Haven't environmentalists debated the possibility of a runaway greenhouse effect for decades? It seems to come in and out of vogue every ten years or so and then they turn the other way and decide that the Earth is heading for a period of glaciation. How certain are you of your findings?"

"Very certain. We have very little time."

"Are you telling me that the world is coming to an end?"

"In a sense, it is. Given the rise in temperatures, the rise in sea levels, the instability in weather patterns and the resultant droughts and famines, a large percentage of the population will die within fifty years and that civilization as we know it will cease to exist shortly thereafter. If it was just the environmental factors, we might last longer; however, the human factor shortens the amount of time that we have. As resources diminish and conditions deteriorate, wars will erupt."

"But there's got to be something we can do? I just can't believe that we are totally impotent." Sam smiled. "After last night and this morning, I know that you aren't. Besides, when you go public somebody will come up with something. They've got to." Sam noticed Jason glancing down at his feet. "What is it?"

"I'm afraid that we can't go public."

Sam looked at Jason questioningly. "Why not?"

"The computer models show that if this information leaks out, the net result will be an acceleration of the greenhouse effect. People will overreact." Jason paused to let the idea sink in. "Think about it. The models are right. Keeping the information quiet is our only option. Besides, I think I came up with a solution to salvage at least a portion of our technological civilization. We can build colonies at the Earth's Lagrange points. We should be able to construct enough colonies to support a self-sufficient population of several million people. That's the only 'solution' that we've come up with."

Sam sat back in her chair, digesting what Jason had told her. Jason looked at Sam sitting back in her chair with her long, wet sandy blond hair draped across one shoulder. She didn't have any makeup on; Jason didn't think she needed any. Her complexion was dark for a blond; she never had the milky white cast about her skin that had been so in vogue during the latter part of the '90s. Her eyes varied in color from blue-green to hazel. Jason shuddered when he pictured Sam's face scarred by the harsh ultraviolet light and her beautiful eyes blinded by cataracts.

"You know, I've always been fascinated with the space colony concept. Gerald O'Neill, the originator of the idea of constructing colonies at the Earth's Lagrange points, was a professor at Princeton University at the same time as Geoff Barnes' father. I sat in on a conversation that Geoff and his son Jimmy were having at their house in Princeton one day last year. Jimmy was studying the concept in a class at Lawrenceville and mentioned it to his father. Geoff went into some detail about his meetings with the late Professor O'Neill and how his father and Professor O'Neill had been friends. Jimmy ended up writing a term paper about the subject. I have a feeling that you'll find Geoff a willing supporter."

"I hope so. We'll need him. That's the reason that I can't take the day off. Jack scheduled a Project Runaway meeting at the office at 1:00. I've got to get back."

Sam squeezed Jason's hand. "I understand. I also realize how busy we'll be over the next few months dealing with the crisis. Why don't we get married in the spring? By then you should be far enough along with the project to take the time to get married. But for now, I'd like to try to put this out of my mind and spend a last half hour with you before we return to D.C."

McLean, Virginia

When Jack Watson arrived that morning, he headed straight for Michael Barnwell's office. While Michael did not seem to notice Jack's arrival, Jack knew better. Michael liked to throw visitors off balance by either triggering the door to open as they stepped up to it or by ignoring their presence when they arrived. While Jack found Michael's games annoying, he had hired him anyway because of his extraordinary capabilities. His idiosyncrasies didn't matter.

Jack walked over to a table where Michael had installed his own coffeepot and poured himself a cup. He took a sip and scowled. Michael's coffee tasted terrible. It was strong, thick as syrup, and dark. Michael had once explained that he'd gotten used to that type of coffee while in the Air Force and now preferred it. After Jack took his seat, Michael turned to speak with him.

"Good morning, Jack. How are you?"

"I've been better. How about you?"

"Busy." Michael walked over to pour himself a new cup of coffee. He took a sip and held the coffee in his mouth for a minute, seeming to relish its flavor. He then returned to his chair. "This Project Runaway of yours is going to require far more intricate security measures than we ever intended to establish here. I hope you understand that." Michael stared directly at Jack.

"Of course I do. Why do you think I set up our meeting? Once we obtain Congressman Barnes' approval, our budget will increase dramatically. Don't worry about the money. You'll be hiring off-site investigators to work with you on this one?"

"Yes, I'll use the same group that I've dealt with in the past. They'll have no need to know why they are tailing our people. I've also set into motion the monitoring systems for the office telephone system as well as the home and mobile telephones of all of the members of the team. I've programmed the computer to monitor vocal stress levels and terminate any suspicious calls."

"What about bugs in their houses? I'd really prefer not to do it."

"Don't worry, Jack. I'll only take the measures that I deem necessary. Unfortunately, the probability is that at least two members of the team will break down and tell their spouse about the problem within two weeks. So long as their spouses don't take the matter any farther, we don't have anything to worry about. I've ordered psychological profiles of each of the team member's spouses to determine what risk of a leak they present and should have results for you inside of a week."

Jack nodded periodically. "It looks as if you've done your usual thorough job. Now we need to decide exactly what we are going to tell the team members of our steps. They have all been loyal and have never leaked anything before. I hate to erode their trust."

"Don't worry. I know what to do. We'll let them know that we've taken a number of new security steps as a result of the need for absolute secrecy in the matter. I'll tell them that we've hired a security team. That way, they'll be less likely to leak anything. We've got to get them to understand just how serious this is. Each double-black project has its own security team assigned to it. The security teams will be constantly testing them to keep them in line. In a situation like this, everybody will be expendable. Leaks can not — will not be permitted to happen. I know of one instance where a reporter was killed for getting too close to a project. Before sanctioning the killing, the team confirmed the extent of the breach, destroyed the trail followed by the reporter and did it in a manner that convinced a suspicious editor that the reporter had died of natural causes. These guys are pros Jack."

While Jack authorized extreme security measures, his face became several shades lighter by the time Michael finished. "They would really kill?"

"Of course, Jack. We're not playing games. And I will not be in a position to stop them."

"But who will have control over the security team?"

"It depends on who's brought in. No politician will take the position. It's too risky. A professional, former CIA or defense intelligence, with unquestionable loyalty is generally selected. Were I not so closely allied to EnviroCon, I'd be a good candidate. If we play our cards right, we'll have significant input on the decision."

"I still have a problem knowing that somebody on our team could get killed for leaking information."

"Face the facts, Jack. If this thing leaks out, we won't have a chance. Would you let one person, even a friend, screw up the chance to save millions if you had the ability to stop that person?"

Michael again faced Jack down with his gaze. This time Jack's eyes wandered around the room, finally focusing on a spot on his right shoe. He rubbed at it with his left shoe. It didn't come off. He then looked back up at Michael. "I guess you're right. We are in the midst of an extreme crisis. Extreme measures may be necessary. When we meet with the group again at 1:00 to review the various proposals they've come up with; I'll let you discuss security matters with them. And I'll leave what you tell them up to you."

CHAPTER 5

November 13, 2006.
Xingtou Province, China.

General Hong Liu and Minister Ping sipped a cup of tea in the VIP viewing area as technicians completed the final launch preparations for the first manned launch of the Long March V. Both knew what they had at risk; both had no doubt about success. The Long March V mission, and the missions that would follow, would secure China's leadership role throughout the twenty-first century.

Relatively young to hold full membership in the politburo of China's communist party, a politburo dominated by a gerontocracy since the 1970s, they rose to power as the architects of the stunning success of the Chinese economy during the decade of the 1990s. They oversaw the expansion of the free market economy into China and the assimilation of Hong Kong into China's quasi-free market economy. They had accomplished what few Westerners and few of the remaining hard line communists would have thought possible: successfully convert a centrally planned economy to a predominantly market economy without embracing democracy. The politburo of the communist party continued to reign supreme; their word was law.

General Liu and Minister Ping had learned from the mistakes the Soviets had made. They gradually converted China's economy to a modified market economy, which had become the envy of the world. While Russia and the former Soviet Republics grappled with their economies, the United States struggled to reduce the size of its military establishment and Western Europe preoccupied itself with the assimilation of Eastern Europe's economies, China's economy grew at a rate similar to that of Japan's in the late 1960s. The size of China's gross national product now surpassed that of Spain and Italy. Given current growth patterns China's economy would surpass France's and Great Britain's in the early part of the next decade and threatened to catch up with Japan and Germany late in that decade. No longer an economic paper tiger, China had truly become a modern industrialized nation.

General Liu and Minister Ping also knew that China's economic success story could not last forever. As with the slowdown in growth of the Japanese economy, so too would China reach a point where growth would stagnate; recession would set in. General Liu had begun to expand China's space program in the mid-1990s and had been instrumental in convincing a number of former Soviet space scientists to join the Chinese space effort. Together with Minister Ping, General Liu convinced the Politburo to commit heavily to the Chinese space program. Such a commitment would serve the dual purpose of stimulating growth into a new and important area and providing fodder for China's propaganda machine.

Today General Liu and Minister Ping stood awaiting the launch of three Chinese astronauts into low Earth orbit. The astronauts would spend several days inspecting China's first space station; a space station remarkably similar to the old Soviet Mir stations. If the station checked out, two of the three astronauts would remain on board for several months, until replacements arrived. Both General Liu and Minister Ping envisioned this as the beginning of a permanent presence in space for China.

The Countdown stood at T minus 2 hours and five minutes. While Western reporters had not been permitted to the launch site, if the launch succeeded, a press conference would announce China's intentions to the world and would launch a new phase in China's long history.

◆

Despite Sam's confidence in him and in Geoff Barnes's reaction to Project Runaway, Jason was nervous when he arrived at the office. Too much rested on this meeting. He truly believed in his solution, but feared that if he somehow blew his presentation, he would doom the future of mankind.

Jack entered Jason's office at 8:34 a.m. Jack looked at Jason and then at his Rolex. "Let's go grab a bite to eat; that is, if you haven't eaten yet? We've got plenty of time." Jason didn't really feel like eating, but he agreed. Before Jason had a chance to change his mind, Jack grabbed his arm and led him towards the cafeteria. "While I believe we'll convince Congressman Barnes of the dire nature of the problem we face, I'm not sure how he'll react. At least we are getting started." Before Jason could say anything, Jack continued. "There's just too much tension within the team right now. Robert looks as if he's on the verge of a nervous breakdown. Nakamura is more cantankerous than ever. Everybody's tired. People within the firm are starting to wonder if there's a problem. I've managed to put them off, but we've got to do something. You're the only member of the team that doesn't look worn down. I'm glad that you're heading the team."

"Thanks for your confidence, Jack. In fact, the pressure's getting to me too. Today I'm as nervous as when I presented my Masters dissertation to Walter Bennette."

"You don't look it." They arrived at the empty cafeteria. Jack got a cup of the day's special coffee, Mocha Java, and a toasted sesame seed bagel and cream cheese. Jason momentarily eyed the eggs, but decided against it. Instead he grabbed a butter-scotch muffin and a Diet Coke.

They took the corner table. Jack spread the cream cheese on his bagel as Jason took a bite of his muffin. The butterscotch melted in his mouth. Jack broke the silence. "You know, it still amazes me that you completed the program as fast as you did. I hired you because you were one of the top programmers in the country, but I didn't think that anybody could write so many lines of code so quickly."

"It wasn't that big a deal. The Omegatron's the real hero." Jack and Jason finished their breakfasts. Jason felt far more at ease than he had last night or first thing in the morning. It suddenly dawned on him that Jack's purpose had been to calm him down. While still nervous, the butterflies in Jason's stomach had returned to their cocoons.

After leaving the cafeteria, Jason returned to his office to pick up the "secure" briefcase that Michael Barnwell had given him. If Jason failed to open the case correctly, a small canister of hydrofluoric acid would wipe out its contents. While Jason understood the need for security, the "secure" briefcase seemed a bit too cloak and dagger.

They rode in Jack's car, a 2005 silver Cadillac Seville. Jason had to admit that it was far more comfortable than either his Saturn or Sam's Jaguar. The code programmed into the car's computer system enabled Jack to travel in the HOV lanes on the D.C. freeways and to park in the new congressional garage.

When they arrived, Jack led the way up to Congressman Barnes' office. They only waited a few minutes for the Congressman to emerge from his office. Geoff Barnes walked directly over to them with a grin on his face and his hand extended. "It's great to see you both. I always look forward to our meetings."

Jack responded first. "It's nice to see you, Geoff. Thanks for taking the time to meet with us on such short notice."

"Think nothing of it. I know that you wouldn't press for a quick meeting if it weren't important. Besides, my alternative was to sit in on a bi-weekly committee meeting. I used this meeting as an excuse to get Ben Slater to cover for me." Geoff Barnes led the way through his staff offices to his conference room. When they entered, Sam was standing over the computer terminal. She turned to greet them and shook all of their hands. Jason and Sam had both agreed to keep things strictly professional when working together. Sam did, however, give Jason's hand an extra squeeze that brought a smile to Jason's lips.

Geoff Barnes took the seat at the head of the table with Sam to his right and Jack and Jason to his left. Jack began the meeting. "Geoff, before we begin, I want to confirm with you that this room has been secured."

Geoff inclined his head towards Sam. "Sam took care of that this morning." Sam nodded.

"Good. After you hear Jason's briefing, I think that you'll agree with the need for the security sweep." Jack turned towards Jason. "Jason, the floor is yours."

Jason began to describe what they had learned about the runaway greenhouse effect. Half an hour later, he finished. Congressman Barnes did not once interrupt him. After finishing, Jason watched the Congressman for a reaction. He had expected to see shock or disbelief in Congressman Barnes' face, but he saw nothing. Evidently, Geoff's training as a politician had enabled him to mask his reactions.

Finally, Geoff broke the silence. "That's one hell of a presentation. I'd expected something unusual, but this, well, unbelievable. Are you sure your results are accurate?"

"Yes. They're accurate."

Geoff turned to face Jack. "Are you sure about this, Jack?"

"You know that I wouldn't be here if I wasn't."

The Congressman frowned. "Who knows about this?"

"Besides the four of us, the five other members of the Project Runaway team. We've tried to keep this as quiet as possible. In fact, we determined that a news leak will accelerate the pace of the greenhouse effect."

The Congressman leaned back in his chair. "You're probably right. But, frankly, I doubt that very many people would believe you. I myself am having trouble accepting the information; and I know more about the environment than most people and know and respect both of you and your firm. Sure, a few enviro-radicals will jump all over the data and sound their apocalyptic trumpet, but few decision makers will believe you."

"That's true. Unfortunately, once the information is out and the radicals begin sounding their trumpets of doom, additional studies will be conducted. The data will, ultimately, speak for itself. We've estimated that credibility will remain a factor for several years, but, gradually, the public at large will come to believe the data and will react to it. We can't take the chance that our studies are wrong. Jason, where is the table that reflects the release of the information to the public?"

Jason responded immediately. "Table 14 on page fifty-two of the Report." Geoff Barnes took a minute to flip to and review the table.

"We need to conduct additional studies, but your table is perfectly clear. We don't have the option of going public on this one. I personally dislike keeping information from the public, but I am a realist. Occasionally, it's necessary."

Geoff Barnes appeared deep in thought as they waited for him to ask another question. Sam, who knew him better, knew that he might continue to ponder issues in his mind for an inordinately long period of time. While a detriment when dealing with constituents, Geoff's tendency became an asset during negotiations. Sam had observed a number of Senators and Congressmen fidget uncomfortably while waiting for Geoff to make the next point. They rarely outlasted Geoff. For Jason's benefit she intervened.

"Have you come up with a solution to the problem?" Somewhat startled, Geoff Barnes turned his head towards Sam. However, after realizing what he had been doing, he nodded approvingly.

Jack answered. "Sort of. We've got a plan and, while it's not a solution, we believe that it is the only possible course of action." Jack nodded at Jason to take the lead.

"That's right. We believe that the key to the survival of a technologically advanced civilization is to establish a self-sufficient civilization off the surface of the Earth." Jason paused to let his statement sink in. "While building structures capable of surviving the climatic changes here on Earth is easier, quicker and cheaper to accomplish, and while more people could be accommodated, the idea won't work. We can't construct enough facilities to house and feed all of the people currently living in the United States, let alone the entire world. When people realize this, they'll fight for a place in the specially designed communities. The attacks could destroy the communities. Our studies predict that sections of the armed services will revolt in order to assure themselves a place in the communities designed to survive. Civil war could break out."

"Even assuming that several communities survive the climatic and sociological upheavals, those communities will, inevitably, deteriorate into dictatorship. They may not cooperate with other surviving communities and will not be in a position to rebuild the planet. As the planet continues to get hotter and hotter, the communities will suffer a lingering death. It might take decades or, perhaps, centuries, but eventually they'll all die out."

"Space-based communities have a much greater likelihood of surviving and even thriving. Being located at two of the stable orbits in the Earth-Moon system called Lagrange points, they will be insulated from the violence and war that will rock the globe. Only the most determined nations will be able to strike at them and they should be in a position to defend themselves. Given the need for mutual support among the colonies, it's unlikely that they'll fight amongst themselves. And while individual freedoms will probably be less than we are used to now, we should be able to set up checks and balances to avoid outright dictatorships."

"Finally, and most importantly, we believe that a space-based industrial civilization will be in a position to eventually fix the climatic problems that their planet-bound ancestors created. Whether they will choose to do so is another question entirely. But at least they'll have the option. Congressman – Geoff, we can't do this without you." Jason looked the Congressman straight in the eye.

After a brief pause, Congressman Barnes responded. "Thank you, Jason. You've given me a great deal to think about. One thing that you didn't mention; did you consider constructing facilities under the oceans?"

Jason and Jack looked at each other. "In fact, we, Jeff Nakamura, did. We studied the idea, but decided against it. Undersea cities of the scale we would need present almost as many technological challenges as building cities in space. In addition, given the increases in storm activity and the predicted rise in sea level, we don't have time to construct enough underwater cities to make a difference. We also don't know enough about the effect of the climatic catastrophe on ocean life to determine whether the cities would be able to raise enough food to feed themselves and make them independent from the surface."

"Have you thought about how to implement your program? After all, the cost of the program will be astronomical."

Jason and Jack had agreed before the meeting that Jason would answer any "technical" questions and that Jack would address the "political" questions. Jack responded, "Actually, while we do have a few ideas, we wanted to consult with you before coming to any conclusions. Given the need for tight security, we initially thought about funding the program as a 'double-black' project."

Geoff Barnes' eyebrows rose up upon Jack's use of the term "double-black." "Just what do you mean by a 'double-black' project?"

"I wasn't sure whether you had ever heard of one, Geoff. We all know about the "black" sections of the Pentagon's budget. Only the President, a few staffers and one or two Congressmen from each party know what the 'black' elements of the budget are spent on. During my years in Washington I have heard rumors of projects that are funded without the knowledge of the President or of Congress. But, until our Security Chief indicated that he had actual knowledge of such a project, I assumed that rumors were all that existed."

"That's enough, Jack. I know all about them. It's a good thing that this room is sealed. Very few people in Washington are certain of their existence. I haven't even discussed double-black projects with Sam before. They do exist. I know of several. In fact, one is currently up for review that might just come in handy. Needless to say, you will all keep what I am about to disclose to you strictly confidential. I don't even want you to share this information with the other members of the team. At least not yet. Eventually we'll be able to incorporate and expand the existing program into a broader program for Project Runaway, but, for the time being, keep it in strict confidence."

A buzzer rang, interrupting the Congressman. Everybody looked up, startled. Jason felt a little silly, but he remembered what Michael Barnwell had said about double-black projects. He had felt, at least for a moment, that they'd been busted. Sam got up to open the door. She explained: "The room is soundproof. The only way to 'knock' is to ring the bell." Jason remembered the screen on the wall and blanked out the image just before Sam opened the door.

As she opened the door and a youngish man dressed in grey flannel pants, a white button-down pinpoint oxford shirt and a striped tie rushed into the room, short of breath. The Congressman's secretary came in right after him. "I'm sorry, Congressman. I told him you were not to be disturbed. He didn't listen."

"That's all right, Dawn. Knowing Peter, you did the best you could. Just go back to your desk and close the door behind you. If anybody asks, I'll be in conference for another hour."

"Yes sir." Dawn left, closing the door behind her.

"Now, Peter, what's so important that you had to interrupt us?"

"I'm sorry, sir. But the Chinese just launched the Long March V. I knew that you'd want to see this." Peter pressed several buttons and CNN news appeared on the screen that Jason had been using to illustrate his presentation. A large picture of Bradley Nesbitt, CNN's anchor, filled the screen.

". . . live for the press conference at the Beijing Space Studies Institute. Karen, are you there?" "I am, Bradley. While we are waiting for the press conference to begin, I'll repeat what official Chinese news sources just told me. The Chinese have just launched their first manned space venture. Three astronauts are on board the capsule. They have just docked with the Golden Dragon, the first Chinese manned space station." The reporter stopped for a second and pressed her hand against her earpiece. "Ah, here we are. Quang Lai', spokesman for the Chinese Space Agency, is at the podium."

"Ladies and gentlemen of the world press, the people of China are proud to announce the successful launch of three Chinese astronauts into space. As we speak, the first of our astronauts is entering the Golden Dragon for a six-month stay. From now on the people of China intend to maintain a permanent presence in space. Over the past decade, China has evolved into an economic superpower. With the conquest of space at hand, China will maintain its role as a world leader throughout the century and will be at the forefront of the development of the resources that space has to offer. . ."

Congressman Barnes walked over to the console and hit a button, turning down the volume on the screen. "Peter, you were right to interrupt me. This is a very significant development. Unfortunately I don't have time to watch the entire telecast. I'd like you to prepare a summary intelligence report for me. The press is going to be all over this. It won't be long before they start knocking on our door for an interview. It's up to you to get me up to speed and keep me there." Congressman Barnes walked Peter to the door and closed it after he left. "Now, where were we?"

"You were about to discuss an existing double-black project."

The Congressman nodded and returned to his seat. "As far as the public knows and, for that matter, most of the members of

Congress and the President, we've never been back to the Moon. Our space program has dwindled to a shadow of its former glory. However, just prior to the end of the Cold War, President Reagan, several top officials at the Pentagon and several key members of Congress decided that we needed to accelerate the Strategic Defense Initiative. They established a command base on the Moon codenamed Project Starcom. Before a subsequent President scaled the project back, we had established a temporary base on the Moon and stockpiled it with supplies, including a small nuclear reactor. Several Air Force astronauts traveled to the base and spent a brief period of time on the Moon before being ordered to return."

"About a year ago a retiring member of Congress brought me into the loop as his replacement on the oversight committee. We had planned to mothball the program this year. It shouldn't be too difficult to convince the other members of the committee to reallocate the assets to a new project. So we should have a source for initial funding of the project and a head start that I seriously doubt you'd considered."

"That's great news. It'll save us a year, at least. We'll jumpstart the development of the mining facilities and mass driver. But I can't believe that we've been back to the Moon and the public doesn't know about it."

"It's not that hard to believe. Think about it. Before the end of the Cold War we spent hundreds of billions on defense. Cash flowed freely into the Pentagon's budget. It was easy to hide the funding for the program in the numerous "overruns" that various high ticket and visible programs suffered. In actual fact, those "overruns" were far smaller than the Congress and the public were led to believe. I still find it hard to believe that most of the members of Congress, the public at large and the press actually believed that the Pentagon spent $500 for a $10 wrench. But they did. The deception worked well, funding several double-black programs. In fact several dedicated officers actually sacrificed their careers in order to keep the secret of the cost overruns."

"The lunar base program will give us the head start we desperately need. This somewhat unexpected announcement from the Chinese will breathe additional life into your proposal, Jason. The public has become increasingly suspicious of the Japanese and the Chinese over the past few years. If we play this right, we can reinvigorate the flagging U.S. space program. Your project will have both a public and a hidden source of funding. We'll be in business. But before we go any farther, I'd like to review your Report and set up a meeting with Gene Samuelson. Will you both be available over the next couple of days for a meeting?"

"Of course. We'll be available at any time, day or night."

"Good. Thank you gentlemen. Sam will call you to schedule the meeting." Sam showed Jack and Jason the way out. Jason started to say something to Jack in the corridor, but Jack signaled him to keep quiet. Only after leaving the Capitol Hill area behind did Jack say anything to Jason about the project.

"I'm sorry to have cut you off, but with a project like this we can only talk in secure areas. In fact, we shouldn't discuss anything about the Project until we get back to the office. While I had Michael sweep the car for bugs earlier, one could have been planted while the car was parked in the garage."

Jason looked over at Jack. His hands squeezed the wheel and he kept looking into his rear view mirror. "Are you serious, Jack? After all, nobody has any idea about what we are doing. Nothing could have leaked."

"That's probably true, but espionage of all sorts remains a problem in our business. In fact, on a past project, we found a bug in Charles Lane's car. In addition, people do, occasionally, watch our actions just to pick up information that might turn out to be of value to somebody. While such incidents are rare, we can't afford to take a chance. Our sudden, unscheduled, meeting with Congressman Barnes might pique somebody's interest. Hopefully, Barnwell's security teams will sniff out any random surveillance before we give anything away. Let's not say anything more. If somebody was listening, just talking about the need for security is giving away too much information."

"That sounds a bit paranoid, Jack."

"It pays to be paranoid in this business. Now, about that wedding of yours, have you and Sam set a date?" They discussed everything but the Project the rest of the way back to their office. When they arrived, they found a message from Natalie at the reception desk. She needed to see them as soon as possible. They headed straight for her office. She wasn't there. They then went to Jason's office. When they opened the door, they found Natalie pacing in front of the sofa, smoking a cigarette. Her hair was disheveled, her eyes, bloodshot and her cheeks, stained with tears.

"Natalie, what's wrong?"

Between sobs Natalie answered them. "My father called. It's my mother. She hadn't been feeling very good, but refused to go to the doctor. Last night she began to throw up blood. They went to the hospital. It's cancer."

Jack and Jason led Natalie to the sofa and sat her down. Jack pulled a white handkerchief out of his pocket and handed it to her. "Is there anything that we can do?"

"No, but I'd like to go to her. The doctors told my father that she only has a few weeks."

Jack answered her. "By all means, go. You should have already left. You didn't need to wait to ask."

"But the project?"

"We'll manage without you. Don't think about it. Take a few weeks to be with her."

"Thanks, Jack. I'll leave right away."

"Call us when you have any word or if we can do anything for you."

"Thanks. I'll keep you posted." Natalie wiped away the last of her tears with Jack's handkerchief and handed it back to him.

CHAPTER 6

Coach Joe Kalvanovich is concerned over the prospects of the upcoming game with the AFC east division leading Miami Dolphins. The Redskins are neck and neck with the Atlanta Falcons for the lead in the NFC East midway through the season. Looking at the remaining schedules, this game could be crucial to deciding the winner of the Conference. Atlanta beat Miami in the second game of the season, the only game lost by the Dolphins. Atlanta is facing the lowly Colts and a win appears certain. To remain tied with the Falcons, Washington has to beat Miami. In the unlikely event that Washington and Atlanta tie for first place in the division, one of the tiebreakers used is the results of games against common opponents.

With the loss of their speedy wide receiver, Jefferson Gaines, last week in the overtime win against Philadelphia and given the record shattering 103 degree heat projected for the game in Miami, Coach Kalvanovich is concerned how his remaining receivers will fare. The heat, the tough Miami defense, and the lack of relief for the receivers may prove to be too much. The line on the game is Miami minus 5 points.

✦

McLEAN, VIRGINIA.

Jason had just come back from lunch and was finishing reading the paper with his feet propped up on his desk when Jack Watson walked into his office. "You ready to go." They were scheduled to meet with Congressman Barnes and Eugene Samuelson, the National Science Advisor, at the Old Executive Office Building in just over an hour.

Jason slid his feet back down to the floor. "Don't let my relaxed appearance fool you. I'm ready. I'm just waiting for Michael to run a few more copies of the Report. Any hint from the Congressman on what to expect?"

"No, just the message about the meeting. And get five copies, one for you, one for me and three for the Congressman. By the way, we'll take my car. Michael installed some additional security equipment in it."

"Good. I'd just as soon not have to drive in downtown traffic. I'll meet you at the reception desk." Jack nodded and walked out.

The trip downtown was uneventful. While a bit nervous, Jason wasn't stressed out like he'd been prior to the first meeting. In fact he was looking forward to this meeting. For the past few days, he'd been in limbo, not knowing where things were going. One way or another, today's meeting would move events forward.

Jack pulled into the driveway of the underground garage. A security guard opened the gate. They parked and then went to the main reception area where they had to wait for a few minutes. A security guard appeared to escort them to the conference room.

They walked down several brightly-lit corridors to a bank of elevators. It looked the same as the building that housed Congressman Barnes' office. Were all government office buildings constructed like this? When the elevator finally arrived, they took it to the third floor and got out. They passed along several nondescript corridors, making several turns. Jason was surprised as to how few people he saw in the halls. They finally arrived at a door at the end of one corridor. Their guide opened the door for them. Congressman Barnes and Sam stood at one end of the table chatting with a tall, lean grey-haired man in an ill-fitting suit. Jason recognized Eugene Samuelson. While they had never met, Jason had once sat in on a lecture that Professor Samuelson had given at MIT. Although not a brilliant particle theorist, Samuelson had a reputation as a cagey bureaucrat. Before his appointment as National Science Advisor, he had acted as deputy administrator for the Texas Super Collider project.

Congressman Barnes saw them enter. With a wave of his arm, he signaled for them to come over. "Gentlemen, nice to see you again so soon." They shook hands. "I'd like to introduce both of you to Gene Samuelson." They exchanged greetings with Samuelson and Sam. Jason had to restrain himself from giving Sam a kiss. "Are we ready to get started?"

"Not quite. We're expecting one more person in a few minutes. Sam, why don't you help Jason set up while we wait?" By the time Jason and Sam had finished setting up, a door in the back of the conference room opened. A young man entered and began scanning the room. His eyes didn't miss a detail. He wore a navy blue suit, white shirt and a very conservative yellow tie. He whispered something and then left the room.

Shortly thereafter a man in a grey suit entered the room. Immediately behind him a figure entered that Jason recognized instantly; Vice-President Tom Hammond. Jason had read that Hammond had been ostracized by the Administration and had no "real" duties. The Party had added him to the ticket at the national convention as a concession to party moderates. While he had campaigned hard and, according to the political pundits, successfully for President Wright, they had a falling out shortly after the inauguration. President Wright had relegated the Vice-President to head up a number of unimportant committees and councils.

The man in the grey suit stood to one side and scanned the room. It took Jason a minute to realize that the man carried himself like Michael Barnwell. The Vice-President turned towards him. "George, if you don't mind, please wait outside." The man nodded and left. Jason surmised that both men were a part of the Vice-President's Secret Service detail.

The Vice-President first walked over to Eugene Samuelson, who was standing closest to him. He greeted him and then greeted Congressman Barnes. They spoke briefly and the Vice-President laughed and patted Geoff Barnes on the back. Jason recalled that Tom Hammond and Geoff Barnes had entered the House together in the same freshman class. The Vice-President then walked over to Jack Watson and introduced himself and finally approached Jason and Sam. He said hello to Sam first. They had evidently met before; Jason planned to ask Sam about it later — and then introduced himself to Jason.

"Mr. Graham, it's a pleasure to meet you. I've heard good things about you." He extended his hand and Jason extended his own for a firm handshake. Jason doubted that the Vice-President knew anything about him and was just being a typical politician, but he couldn't be sure.

"I'm honored to meet you, sir."

The Vice-President turned back to where the others were standing. "Why don't we get started? Unfortunately, and despite what the press claims, I have a tight schedule. I can only give you gentlemen half an hour." The Vice-President took the seat at the head of the table. Jack nodded to Jason and took a seat next to the Vice-President. The National Science Advisor sat down next to him and Congressman Barnes next to him. Jason handed Sam four copies of the Report.

"I've handed Ms. Whitlock copies of the same Report that we delivered to the Congressman at our briefing two days ago. The Report cannot be copied. The paper is treated so that it will turn black if it is copied or photographed with a flash. If any of you need additional copies, please let either Jack or me know. We'll get them for you."

The Vice-President began to leaf through the Report. The National Science Advisor followed suit. Jason cleared his throat.

"If you don't mind, you've only given us half an hour. You can review the Report later." Both men stopped leafing through the Report, the National Science Advisor's face reddening. However, the Vice-President did not appear at all upset by Jason's mild rebuke. "Sorry, Mr. Graham. Bad habit of mine. Please continue."

"Since I don't know what information Congressman Barnes already discussed with you, I'll start at the beginning. My firm received a contract to write a program to categorize and help analyze the vast quantities of data that NASA satellites have been gathering for years. We completed the project. The results we obtained were disturbing, to say the least. Gentlemen, we have a very high degree of confidence that the Earth's climate is in the early stages of a runaway greenhouse effect. Unfortunately, unlike most climatic changes, this change is occurring very rapidly. We project that the effects of the process will become clearly noticeable in five to ten years regardless of anything that we presently do and that the planet will be rendered virtually uninhabitable within 50 to 150 years."

"We haven't run our projections out beyond that point to determine when the planet will reach a new equilibrium. Although it's doubtful that the Earth will become another Venus, we are in serious trouble." Jason used the remote to turn down the lights in the room and to illuminate the first of the charts on the large screen behind him.

"As you can see, global temperatures have risen dramatically over the past several decades, on the order of several degrees centigrade. The increases will become even more dramatic over the next several decades. I won't bother you with the details of how we came to our conclusions. You will find those details in the Report. While all areas of the globe will feel the effects of the rise in temperature, coastal regions will get hit first and will get hit hardest. Sea levels are beginning a rise that will accelerate throughout the next two decades. As ocean temperatures rise, hurricanes and typhoons will pound coastal regions. But interior regions will not go unscathed. The instability in the atmosphere will cause severe heat waves, droughts and floods.

Food production will drop dramatically as weather patterns become far less predictable."

"Basically, gentlemen, if we do nothing, civilization as we now know it will cease to exist within the next fifty years. Mankind itself might fail to survive. If nothing is done we will witness the largest mass extinction of plant and animal species this planet has even known." Jason paused for a minute to take a sip from the glass of water that Sam had placed in front of him. Jason glanced at his watch.

"We have also determined that going public with this information will only accelerate the decline." Jason changed the display on the screen. "The only variable in the two charts is the date that the public becomes aware of the fate of the Earth's climate. As you can see, the downfall of civilization is accelerated significantly by the revelation of the information. That is why we urge that you place the highest possible security classification on Project Runaway."

"Before I finish, the one thing we haven't discussed is what we propose to do about the problem. The only workable solution we have is to construct a series of space cities at the Earth's Lagrange points; stable orbits in the Earth-Moon system. The cities will be independent of the need for support from the Earth and will be capable of maintaining a technologically advanced and advancing civilization. If we obtain the necessary resources and commitments, we can establish a population living on space cities on the order of several million before things deteriorate too far on the planet. To do this we propose setting up a 'double-black' funding program for Project Runaway to develop the technologies and infrastructure that we will need to implement our proposal."

"We have a great deal of work to do refining and implementing our proposals, but we do have a workable program. Thank you for your time. If you have any questions now or after you review the Report, please call either Jack or myself. We will be happy to work around your schedules to get back with you. Any questions?"

Eugene Samuelson cleared his throat as if to ask a question, but the Vice-President held up his arm, cutting the National Science Advisor off. "I'm sure you have a number of questions Gene, but I've only got a few minutes. I have a few observations to make and then I've got to leave. After I go you can continue the meeting and I'll get Gene to brief me later."

"Mr. Graham, thank you for your Report. Congressman Barnes didn't exaggerate when he told me of the urgency of this meeting or of your capabilities. I'm sure that I'll find the Report thorough. However, before you or your firm do anything else or talk with anybody else, I'd like to have your data checked. With something like this, we've got to be certain. Gene and I will determine what other group will review your data. After they report to us, we'll get back with you. I'd like you to pledge your full cooperation with the new group."

Jason nodded.

"Good. Now, the check of your data will take some time. I understand the need to move on this. Rather than having you wait, I've got several matters that you can be working on. First, your report is due out shortly; isn't it?" Jack responded.

"That's right."

"Can you accomplish a delay?"

"Yes; especially with the support of Congressman Barnes."

"Good. Geoff, on my personal authorization, I'd like EnviroCon to work on falsifying the data in the NASA data banks so that independent investigators will not be able to come to the same conclusions that EnviroCon did. Can you do it?"

Jason and Jack looked at each other. Jack answered. "Yes sir, Mr. Vice-President. It will take time, but I believe that we can do it."

"Good. Geoff, I'd like you to increase the funding to EnviroCon so that they can expand their work on the project."

The Congressman nodded. "A small, say thirty to forty percent increase, shouldn't be a problem. I'll call Chris Cadenhead at NASA to obtain his cooperation."

The Vice-President turned his gaze back to Jason. "Next, I'll work on setting up the double-black funding elements of the program. Assuming that the other group confirms your results, we will need to move forward quickly. Geoff and I have agreed to continue the operation of the other double-black project you discussed and to change its focus towards Project Runaway. Congressman Joel Friedman heads up that project. I'm sure he'll cooperate with us."

The National Science Advisor interrupted. "But, Mr. Vice-President, don't you think we'll need the President's authorization?"

"We'll have to travel without it, Gene. It would be a major mistake to reveal anything to the President. The information would be out in a matter of weeks. We simply cannot afford that. Besides, this "matter" falls within my authority as head of the National Space Council. I can authorize it without presidential action. And, unlike the President, I am willing to take responsibility for my actions."

"Finally, under my authority, I'm classifying your data and Project Runaway as top secret and of vital interest to the Government of the United States. You will be guilty of a federal offense if you reveal anything about Project Runaway to anybody without my specific authorization. I hope all of you understand what that means." The room remained silent.

"Good. I'll be in touch. In the interim, keep me posted." The Vice-President got up to leave. Everyone else in the room stood as the Vice-President walked out of the door through which he had entered. Jason thought to himself that it was too bad that Tom Hammond wasn't the President. But then again, he doubted that not being the President would slow Hammond down much.

After the Vice-President left, Jack began to speak with Congressman Barnes. "Now I know why you wanted him to come to the meeting, Geoff."

"I know. I only wish that we belonged to the same party. I'm afraid that I may have to run against him at some point in the future."

Geoff Barnes turned towards the National Science Advisor. "Gene, I can only stay a few more minutes, but I'm sure that

Jack and Jason can remain as long as you like. Do you have any questions for them?"

"In fact, I do. I'd like to go over a number of details with them now and more after I read their report. But before we get to that, are you sure that we can do this without presidential authorization?"

"Yes, I'm sure. The Vice-President does have the authority to initiate a 'double-black' program without the consent or knowledge of the President. Assuming that we move forward, as I'm sure we will once the EnviroCon data is checked out, we'll prepare all of the necessary orders and directives to keep our actions legal. In the interim, Gene, I'm sure that you understand the need to keep this from the President."

Gene's eyebrow lifted. "I do."

"I've got to get to another appointment. Sam, why don't you stay here and act as my representative? I'll see you later this afternoon. Jack, Jason, thanks for coming on such short notice. I'm sure we'll be in close touch over the next few weeks."

Congressman Barnes left the conference room. The National Science Advisor spent the next hour interrogating Jack and Jason. At the end of the meeting he appeared satisfied with their answers.

CHAPTER 7

Super Typhoon Eugenia, the strongest storm to hit the Philippines in over a century, left the northernmost coast of Luzon reeling. Sustained winds in excess of 190 miles an hour ripped apart entire villages. The government in Manila has issued pleas for international assistance. Due to continued high seas and the destruction of most roads entering the region, neither officials nor relief workers have reached the hardest hit areas. Tens of thousands are presumed dead and hundreds of thousands homeless due to coastal flooding, interior mudslides, and the high winds. This is the worst natural disaster to hit the Philippines in a century.

The storm, while weakened, continues to have sustained winds in excess of 100 miles per hour and is anticipated to regroup and gain additional strength before its next projected landfall along the southern coast of Mainland China.

✦

CHEVY CHASE, MARYLAND.

Natalie left her mother's room at the Shady Grove Adventist Hospital after having spent a restless night. Her mother had barely recognized her through the cloud caused by the pain and the medication. She had called Natalie her little Natushka, something she hadn't done since Natalie had been a little girl, and had rambled on in Russian. Natalie could hardly understand her.

Her father, beside himself with grief, refused to leave her mother's bedside. When Natalie got to the waiting room, she sat down and pulled a handkerchief out of her purse. She wiped her face off and then proceeded to pull out her compact in a vain attempt to fix her makeup. A gentleman sat down several seats away from her and lit a cigarette. Dressed in a herringbone tweed jacket and grey pants that matched the greying hair on his head, he looked vaguely familiar; clearly Slavic. Natalie continued to glance at him periodically, but couldn't place him.

She dug through her purse looking for a cigarette, but couldn't find one. She walked over to the gentleman. "Excuse me. Do you have an extra?"

"Certainly." The man spoke with a slight British accent. He pulled a soft pack of Marlboros out of his jacket pocket, shook the pack so several cigarettes stuck out the end, and offered one to Natalie. She drew one out, hesitated for a minute as she noticed the scars on his hands, and put it in her mouth. The man drew out a gold lighter and lit her cigarette.

"Thank you." Natalie drew in a lung full of smoke and barely managed to suppress a cough.

"You're not sick; are you?" The man asked in a concerned voice.

"Oh, no. Just tired."

"You're visiting somebody here in the hospital." Natalie nodded her head as she drew in another, smaller puff from the cigarette. They both sat in silence while smoking their cigarettes. He put his out in the ashtray and walked over to the nurses' station. Natalie finished a minute later, put her cigarette out and began to head back to her mother's room. However, while walking past the man, she overheard him ask for Mrs.

Ivanenko's room. Natalie stopped abruptly.

"Excuse me. I didn't mean to eavesdrop, but did you ask for Mrs. Ivanenko?"

"Why, yes. I did."

"I'm Natalie Ivanenko. Can I help you?"

The man, somewhat startled, squinted his eyes to take a close look at her. "Yes, I see it now. You look a great deal like your mother did when she was younger. We've met before. You were about this high." The man extended his palm and held it just over his knee. "I should have guessed who you were. Little Natasha. Of course." The man's face broadened into a grin. Natalie waited for an explanation, but the man remained silent.

"I'm sorry. But I don't remember your name?"

"Of course you don't! I apologize. Vladimir Kreschenko." He extended his hand. Natalie took it. His palm felt rough, his grip, firm.

Natalie searched her memory. She vaguely remembered her father talking about Volodia Kreschenko. "I do remember my parents mentioning your name a few times. But not recently."

"I've been out of the country for years. I just returned six months ago. By the way, I'm really sorry to hear about your mother. We were very close years ago. How is she doing?"

Natalie's eyes began to water, but she managed to hold back her tears. "Not very well. The doctors don't give her very long."

"How's your father?"

"Rough. He's taking it real hard."

"Yes, I can imagine. He always was a very emotional man; very Russian."

Natalie had never really thought of her parents as Russian or of her father as emotional. They had emigrated to the United States at the end of the Brezhnev era. Natalie had been born in upstate New York where a number of Russian émigrés lived; however, shortly after her birth, her parents had moved to Maryland. Her parents had never talked about Russia. Natalie looked quizzically at the man.

"Can I come see your mother?"

Natalie paused. "Let me check with my father. The doctor said no visitors, but I'll check."

"That's fine. Thank you, Natasha."

Natalie walked down the sterile corridor to her mother's room. Natalie hated hospitals; the smells, the looks, everything. When she got to her mother's room, she found her mother sleeping and her father sitting in the chair next to her, holding her hand and mindlessly watching an old game show on television.

"Papa, why don't you go to the room and rest? I'll stay with mama."

He looked up at her, his face drawn and his eyes hollow. He had not slept more than a few minutes at a time for the past few days. "No, I've got to stay. I've got to be with her."

"Okay, Papa. But at least try to get some rest. Mamma's resting." Natalie touched her mother's hand.

"Fine, dear. I'll try. This chair is mighty comfortable." Mr. Ivanenko leaned back hard in the chair and, with a slight jar, it reclined. He didn't look comfortable. "You go on over to the motel. You haven't slept for a long time either; have you?" Natalie suddenly felt very tired. Her father was right. Between the stress from work and now, her mother, Natalie was exhausted.

"All right, Papa. But if anything changes, give me a call." Natalie leaned over the bed and kissed her mother's cheek or what little remained of it. She then walked over and kissed her father's cheek. She could taste the salt from his tears. Just before walking out she remembered Vladimir Kreschenko. "I almost forgot. I ran into an old friend of yours and mamma's in the lobby, Vladimir Kreschenko. He asked if he could come up to see mamma."

Her father's response shocked her. His ashen face flushed and his bloodshot eyes burned. "Who did you say?" He bellowed so loudly that his wife stirred in her bed.

Natalie spoke very softly in response. "Vladimir Kreschenko."

"Damn it!" Then, seeming to gain control of himself once again, he said, "Just tell him no. Your mother is not up to visitors. Please give him my regards, but tell him no."

"Is there anything wrong, Papa?"

Mr. Ivanenko shook his head. "I'll let your mother know that he came by. They were close once. But please, don't bring up his name again." Her father slumped back into the recliner, suddenly looking very old.

"All right, Papa." Natalie reluctantly backed out of the room. She knew her father was holding back. She'd never seen him react like that. His sudden anger didn't make sense. Granted, his defenses were down; but, even so, it was too uncharacteristic. Natalie decided to try to get to the bottom of the situation.

When Natalie returned to the lobby, Vladimir was waiting for her. He had just bought a bunch of balloons to send up to her mother. Natalie walked up behind him as he was leaving the balloons with the nurse. "Mr. Kreschenko . . ."

"Ah, you're back."

"Yes." Natalie's head drooped slightly, as if she were embarrassed. "I'm sorry, but I'm afraid that you can't see my mother. Doctor's orders."

"But it is I that should be sorry. There's no reason to apologize. Believe me, I understand." They stood for a moment, looking at each other awkwardly until Mr. Kreschenko glanced at his watch. "It's almost six o'clock. Have you had dinner yet?"

Natalie was tired and didn't feel like eating, but decided that this was a good time to try to decipher why her father had reacted so to this man's name. "Not yet. I'd planned to go back to my father's room and get some rest." Natalie paused. Mr. Kreschenko looked dejected. "But I wouldn't mind having dinner with you. I'd love to hear some of the old stories that you must have about my mother and father."

"Good! Let's go." Mr. Kreschenko extended his arm towards Natalie.

"I do feel awfully dirty. Do you mind if I take time to freshen up?"

"Not at all, Natalie. I'll pick you up in front of your father's hotel at 6:45. Will that give you enough time?"

"That'll be fine." They separated when they exited the hospital.

Forty-five minutes later, Natalie walked out of the front entrance to her father's hotel. She felt far better, having showered. She'd also called the hospital and confirmed that everything was okay. The head nurse said that they didn't anticipate any changes for several days.

She looked around. Kreschenko got out of a blue Ford four door sedan and walk toward her. She walked forward to meet him. "I brought my car, even though the restaurant is within walking distance. I thought that you would probably like to stay close to the hospital."

"Thank you. I appreciate your thoughtfulness."

Kreschenko opened the door for Natalie and then got in the car himself. They drove for a few minutes in silence. Kreschenko pulled up to an awning with Marianna's written across it. They left the car to be parked by the valet. Evidently, they were expected because as they entered, the maitre d' led them directly to a small table in an isolated corner of the restaurant. The maitre d' handed them a menu and began to turn away. However, Kreschenko said something to him in French and the maitre d' stopped and turned back. They spoke for a minute and then the maitre d' left.

"I hope that you don't mind. I took the liberty of ordering a bottle of red wine." Natalie shook her head.

"No, not at all." Natalie opened her menu and began to peruse the entrees. She rarely ate in restaurants like this and was shocked by the prices. Evidently, Kreschenko noticed her concern. "By the way, just so you know, the evening is on me."

"No, I couldn't let you pay for me."

"Don't be ridiculous. I invited you out and selected the restaurant. Besides, it is the least I can do for you. After all these years, it's my pleasure."

Their waiter appeared and took their orders. Natalie became uncomfortable as a result of the prolonged silence following their having ordered. She decided to begin her inquiry about Kreschenko's past with her parents. "So, um, Mr. Kreschenko,"

"No, Natalie. Please call me Volodia. I insist."

"Okay, Volodia. I've heard your name before but know nothing about you. Where do you know my parents from?"

Kreschenko pulled a cigarette case out of his pocket and offered one to Natalie. After a moment's hesitation, she took one. He took one himself, pulled out his gold lighter, lit Natalie's cigarette and then his own. He took a long draw from his cigarette and then began to speak.

"I met your mother a long time ago; we were students together in Leningrad. Your mother was a beautiful woman. I fell madly in love with her. Unfortunately, the feeling was not mutual. We got along, but your mother remained oblivious to my feelings." Natalie thought to herself, perhaps that was why her father had reacted the way he had. "To some extent it may have been my fault. I never expressed my feelings toward her openly. I had the opportunity and we became good friends, but before I got the nerve to take things farther, your mother met and fell in love with your father. He had recently moved from Moscow when his father, a minor party official, had been transferred to Leningrad."

"All three of us got along well and remained good friends. Our paths diverged shortly after we graduated. I transferred to intelligence school in Moscow and your parents transferred to Leningrad University. They were good students and, I know you may find it hard to believe, good communists as well. They married shortly after graduating from the University. I didn't see them for several years and missed the wedding; I was away in training. Your mother, however, wrote to me about it. I continued a correspondence with both your mother and your father for years."

"After they graduated from college, your father found a job with the Interior Ministry and your mother with Intourist. Your mother had a gift for languages. However, neither of them was satisfied."

"I, however, was convinced that I had found the perfect opportunity for them. They would move to the United States and serve the State at the same time. I had graduated from intelligence school and had become a junior case officer with the KGB."

Natalie, who had been listening carefully to what Kreschenko had been saying, almost choked on the swallow of water she had just taken. Kreschenko smiled.

"It appears that you are hearing things that you have not heard before. But I'm not surprised. Your parents were perfect. We engineered a conflict between your father and the Party to make him a prime candidate to defect. Your parents played it perfectly. When they finally got the opportunity to travel to Finland, your mother and father defected."

"They moved to the United States and became model émigrés and capitalists. They both got good jobs in New York, your father, with the airlines and your mother as a language teacher. They had you and did nothing to create any suspicions. We knew that the CIA watched all émigrés, but we were very careful. We didn't contact them for five years. By then their covers as dissenters and converted capitalists were well established. I became their case officer after being transferred to the embassy staff. We very carefully maintained contact. They were the best operatives in my stable."

"Unfortunately, just after they had gotten into a position where they might be useful, the Soviet Union ceased to exist. I lost my job and my ability to keep in contact with your parents. They remained in the United States even though their mission had no further meaning. I was not so lucky. I spent many years stuck in St. Petersburg in a dreary job, struggling to make a living."

Kreschenko paused to take a bite from his plate. Natalie, not knowing what to think, looked down and was surprised to see her veal sitting in front of her. She took a bite as well.

"That's why I lost touch with your parents. But now, I have a job with a company similar to your own. It's a think tank funded by several Russian and Ukrainian corporations formed since the fall of the Soviet Union. Private industry finally recognized my expertise in intelligence gathering. Now, I'm back in the United States to gain information for the companies that hired me."

"Natalie, you must realize that your parents loved Mother Russia. They still love it. They were willing to risk detection by the CIA and, possibly, imprisonment for their country. I'm sure that they'll be willing to help out now."

"Natalie, I know that this is going to be hard for you, but Russia, your homeland, needs your help. And we are willing to pay for it. You can take over where your parents left off." Natalie just stared at him in shock.

"Think about it. Russia and the Russian companies I represent need as much help as possible to survive in today's global marketplace. I know that EnviroCon occasionally gets consulting contracts from the government and from private industry. I'm sure that some of the information would be of value to the companies I represent and would help Russia. I know that this is probably a total shock to you and that you will need time to think over my offer. But your help is needed. If your parents were in the position that you are in, they would help me. While the Cold War is over, Russia is struggling with the new trade war. Think about it; will you?"

Natalie sat in silence for a minute. Then she reached for her purse. Kreschenko extended his hand over hers. She yanked her hand free. "Look, Mr. Kreschenko," Natalie emphasized the mister, "there's no way in hell that I'm going to betray my employer or break the law. I respect and like the people that I work with. I will not betray them, either for money or for the 'love' of Mother Russia. Your pleas disgust me. How dare you use my mother's illness as an opportunity to attempt to make me your spy. Even if I believe what you say about my parents and even if they were spies, I am an American and will not betray my country or my company. Now if you will excuse me." Natalie stood up to leave. Kreschenko grabbed her arm.

"Just a moment. Sit back down and lower your voice. You wouldn't want Mr. Wilson over there to hear us; would you?" Kreschenko nodded his head towards Trey Wilson sitting across the room from them. Natalie recognized him. As a reporter for Channel 2 news and a columnist for a national tabloid, Trey

Wilson had written an expose' on Gennady Vorshkoff, the former Soviet spy that he had uncovered living in Maryland. Wilson hounded Vorshkoff until he finally admitted disclosing information to the KGB in the early 1980s. The State Department revoked his citizenship and deported him back to Russia. The story had ended when Vorshkoff shot himself two days before he was to leave for Russia.

She sat back down. "Look, Natalie. I've got to make a living. Your parents were to have been my meal ticket. They have lost their ability to serve me. But there are other ways they can secure my future. That is, if I cannot come up with some useful information for my new employers. I'd rather have you help me out of loyalty to Russia, but..." Kreschenko did not have to finish. Natalie could read in Kreschenko's eyes that he was as cold and hard as a Siberian winter. She didn't doubt that he would carry out his thinly veiled threat. Natalie didn't want to imagine the press hounding her mother during the last weeks of her life or following her father around in his current state. Her parents had done nothing wrong, but she knew what the press and people like Trey Wilson would do for a story.

Kreschenko slipped a card into her hand. "Here's my card. You can reach me at that number at any time, day or night. Think about it." He paused. "But don't think about it for too long. I'm leaving town for St. Petersburg in two weeks. I'll expect a call by then."

Natalie rapidly walked away from Kreschenko. She turned back just before leaving the restaurant and saw, to her horror, Kreschenko walk over and greet Trey Wilson. They appeared to be friends. She turned quickly and left, running the few short blocks to her father's hotel room. She had trouble with the key but finally managed to get in the door, run into the room and collapse on the bed. After a few minutes, she composed herself sufficiently to absorb what Kreschenko had said. She would never have guessed that her parents had been spies. But, then again, what children actually know that much about their parents? She would have liked to discuss her meeting with her father, but didn't want to upset him further. She knew that he would tell her to ignore Kreschenko entirely and would subject himself to the abuses that he would receive from the press in the event that Kreschenko made good on his threat. She couldn't do that. Her father must never know the details of her meeting with Kreschenko.

She thought about calling Michael Barnwell. Natalie had no doubt that Michael would know how to handle a character like Kreschenko — he had been in intelligence all of his adult life — but she feared that Kreschenko might be able to release or sell his information before Michael could stop him. The risk was too great. Natalie knew that it would be hard to keep this from Michael, particularly at night, as she snuggled against him, but she knew she had to. She knew how it would hurt Michael if he ever found out that she had not reported the meeting, let alone reveal any information to Kreschenko, and that, given their relationship, it might put him in a difficult position and hurt his career. This was one decision that she was going to have to make for herself.

Natalie got up from the bed, walked into the bathroom and fixed her hair and makeup. She would go back to sit with her mother and father for the balance of the night and put Kreschenko's threat in the back of her mind as much as possible. Tomorrow; she would make a decision tomorrow.

After walking over to renew his acquaintance with Trey Wilson, Kreschenko returned to his seat to finish his dinner. Unlike Ms. Ivanenko, Kreschenko had not lost his appetite. After finishing his meal, he ate what remained on Natalie's plate as well. He had done well with Natalie. From her expression, Kreschenko knew that he had her and that she had some information of value. When his sources informed him that her bosses, Jack Watson and Jason Graham, had obtained an unscheduled meeting with Congressman Barnes he had hoped that this meeting would be productive. The timely demise of Sonia Ivanenko gave him the perfect opportunity to approach little Natasha. A brief pang of remorse swept over Kreschenko as he pictured Sonia's withering body, she had been a terrific lover, but he promptly dismissed the thought from his mind.

Kreschenko considered himself to have been one of the victims of the collapse of the Soviet Union. His superior had taken him under his wing and Kreschenko felt that power and success were well within his grasp. Unfortunately, his superior had been implicated in an aborted coup against Gorbachev. Kreschenko suffered seriously from the fallout. He lost his job, his apartment and his car and had barely managed to eke out a living for several years. If not for the training he had received in the KGB and his success as a thief, he did not doubt that he would have starved.

But finally, as capitalism gripped Russia, Kreschenko's star began to rise again. He obtained a new master in Nicholai Greschko, a man he respected immensely. Greschko had arisen from the ashes of the Soviet Union to become one of new Russia's great capitalists. Greschko successfully privatized several of the old aircraft manufacturing facilities, converting them into an entity that rivaled the American aerospace giants. Greschko recruited Kreschenko to use the skills that he had learned as a KGB operative to gain knowledge of the operations and secrets of his global competitors. Kreschenko actually preferred working for a private organization. His risks were lower, he had to deal with corporate security departments rather than the CIA and other national intelligence services, and his monetary rewards were greater.

In the Ivanenkos he combined his old world and his new. He enjoyed playing the game of blackmail again. With Little Natasha, he didn't have to do anything special. He had his old case files, his training and his balls; he needed nothing else. Kreschenko smiled to himself as he realized how much he liked his work.

CHAPTER 8

Japanese tuna fleets have reported an unusual number of icebergs in the South Atlantic coming out of the Weddell Sea. One ship barely averted disaster when she struck a berg late one night. While most of the crew evacuated to a nearby Russian trawler, the captain and a small crew managed to keep the ship from sinking.

✦

MCLEAN, VIRGINIA.

Jason arrived later than usual at the office that morning, having spent the weekend at Sam's father's house in Middleburg. The weekend had gone well, better than Jason had expected. Sam's father was extremely pleased with Jason and Sam's announcement and made his pleasure known to all.

Jason figured that part of the reason that Mr. Whitlock focused so much of his attention on Sam was because he felt guilty about how little time he had spent with Sam and her mother while he built his company. Sam had frequently told Jason how much her father had changed since the death of her mother. Jason never would have guessed that Sam had grown up neglected by her father. Mr. Whitlock had come up to MIT every few weeks. He always had some business reason to come to Boston, but Jason and Sam both knew that the real reason was to see Sam. Jason had first met Mr. Whitlock after he and Sam had only been dating for a couple of weeks. Mr. Whitlock had treated both of them to a lavish lobster dinner at one of Boston's four star restaurants.

Jason recalled his surprise when he learned Sam's father's age. In his late seventies, Mr. Whitlock had married Sam's mother after his first wife had died. Her surprise death devastated Mr. Whitlock and shocked him out of his obsession with business. Jason suspected the only reason Mr. Whitlock didn't sell out was his desire to keep the company so that Sam could take it over if she wanted to. In fact, Mr. Whitlock made no bones about his desire to have Sam come to work for his company. Jason suspected that she eventually would. What surprised Jason was the job offer that Mr. Whitlock made to him. It more than doubled Jason's salary, but Jason graciously declined.

One additional advantage that slowing down at work had given Mr. Whitlock was a complete turnaround in his health. Just before his wife died, Mr. Whitlock had suffered his second heart attack. Both Sam and her mother had thought that Mr. Whitlock would die, but he managed to hang on. In fact he looked far better now than he had when Jason had first met him at MIT. He rode horses and occasionally participated in foxhunts.

Mr. Whitlock had inquired about his work on several occasions, both out of curiosity and in an attempt to convince Jason to come to work with him. Jason was glad he could use the cloak of secrecy to avoid discussing work. After his meeting with the Vice-President, Jason had begun the odious task of falsifying the NASA data. Fortunately, he had gotten the able assistance of Robert Wayneright, who clearly needed the work to distract him from his depression, and of Jeff Nakamura, who relished the work for its own sake. Jason had to admit it, Jeff was damn good at the job. As a result, Jason relinquished control of the falsification program to Jeff data and spent his time thinking strategically about their dilemma.

Fortunately, his engagement to Sam could now be made official. It gave him a reason to appear distracted. With her father's approval, he and Sam decided on a short engagement; they would be married in Middleburg at her father's house on March 10. Jason learned from Sam to expect innumerable parties, something that he did not relish.

Natalie cleared her throat, interrupting Jason's thoughts. "Good morning, Jason. How are you?" Jason looked up, somewhat startled.

"Natalie, I didn't expect to see you back so soon. How are you? How's your mother?" Jason looked Natalie over. She appeared rested, but wouldn't meet Jason's eyes. Tears began to well up in Natalie's eyes, but she quickly regained her composure.

"I'm fine, Jason. But my mother's not doing very well. The doctor has given her a few weeks at the most. I don't think she'll ever leave the hospital."

"I'm really sorry to hear that, Natalie. If you ever want to leave to go up and visit with her or if you'd like to take more time off, please do so. And if there's anything that I can do, anything at all, please, let me know."

"Thank you. I really appreciate it, but there's nothing anyone can do." Natalie paused and again averted her eyes from Jason. "My father suggested that I come back to work rather than mope around the hospital. I finally agreed. Frankly, I'll be glad to have something to do to keep my mind off of it. If there are any changes, my father will call me."

"I'm glad you're back. I think that you are probably right to try to go on with your normal routine." Jason chuckled to himself briefly. "Although as I'm sure you'll agree, Project Runaway is anything but normal. In any event, I'll not bring it up again. But if you need anything, even if it is just an ear, or a shoulder, I'll be here. Now, if you'd like, I'll bring you up to speed on the situation."

Jason spent the next fifteen minutes briefing Natalie on the events of the last several weeks. In Jason's view, they were currently in a holding pattern pending further word from Congressman Barnes or the Vice-President. Until they received further direction, Natalie could be a great deal of help in the data falsification effort. Jason suggested that Natalie touch base with Jeff or Robert. Just as Natalie rose to leave, Jack Watson entered the office.

"Jason. . ." He began to say as he entered the room, but then he noticed Natalie rising from the chair in front of Jason's desk. "Natalie, what a pleasant surprise. But..." Before Jack could finish, Natalie interrupted.

"Jack, I think that I know what you are going to say. Jason and I just finished with a similar conversation and I'd rather not go over it again. My mother is dying. Rather than stay at the hospital, I decided to come back to work. I need the distraction. But I really don't feel like talking about it right now." Natalie rose and quickly left the office.

"She seems to be all right, but, I'm not sure."

"I think that she's all right, Jack. She just needs a little time. And, frankly, I agree with her decision. I think that work will help her."

"I hope you're right, Jason. But I'm glad she's back. She should come in handy with the disinformation campaign. By the way, did you tell her to take off whenever she liked to visit her mother?"

"Of course. By the way, I'd rather you not refer to our current project as a disinformation campaign. I don't like the connotations." Jack laughed.

"I know it's not your cup of tea. But don't worry. Nakamura is taking to it real well. With Natalie back you shouldn't have to spend too much time on it. I know that you've got a lot to do with your wedding coming up. Why don't you do it now? You know that once we receive official approval from the Vice-President, you'll have very little free time."

"You're telling me. But I'll be glad when we are finally moving along. I hate this waiting. There's so much that we should be doing."

"I know, Jason. But there's nothing that we can do. Besides, from the call I got today, the wait may not be as long as we'd anticipated." Jason raised his eyebrows. "Geoff wants to schedule a meeting with us for December 28th. He indicated that the Vice-President feels that we'll be able to move forward on that date."

"But why during the holidays, Jack? Not that I was planning to go anywhere, other than to Sam's father's house, but..."

"I know, but it makes sense. This town all but shuts down during the holidays. Congress will not be in session and most of the politicians, as well as their staffs, the lobbyists and the press, will be taking an informal holiday. It will be the easiest time for us to get an entire day with the Vice-President and Congressman Barnes without setting off too many alarms. We don't want anyone to notice this meeting if possible. Personally, it is an inconvenience for me. Michelle had wanted us to go visit her folks in Florida. I'll probably have to fly back early."

"Sorry to hear that, Jack."

"Actually, I won't mind. Her parents are old and just like to sit around and talk all day. I'll be sick of it after a few days anyway. Oh, by the way, are you doing anything this Friday?"

"No. Why?"

"I took the liberty of telling Charles Lane about your engagement. He wants to throw a small party for you and Sam. Is that all right?"

"I think so, but I'll have to confirm that with Sam. I'll let you know later on today."

"Good. See you around." Jack left.

CHAPTER 9

Scientists working at the Peruvian Center for Oceanographic and Climatological Studies issued a warning to all Andean countries and to Central American nations that a strong El Niño is forming in the Pacific. The El Niño will worsen the already harsh effects of the unusually heavy rains that have inundated sections of the Andean countries over the past five years. Ocean temperatures along the coast of South America remain far warmer than normal.

Australian officials anticipate that the drought being suffered in the interior will continue and that coastal regions will need to brace for water shortages. Officials believe they are prepared to handle the forest fires that have broken out in the past in connection with strong El Niño events.

✦

McLean, Virginia.

Jason and Sam arrived at the main entrance to the mansion at

7:15. The party was to have started at 7:00. Jason had tried to hurry Sam, but she insisted that despite the fact that the party was for them, people in Washington never arrived on time. When Jason pulled up to the circular drive, several young men stood along the esplanade in front of the mansion. One opened the door for Sam, while another came around to open Jason's door. Jason hadn't anticipated valet parking. Jack had indicated this would be a small party, primarily for EnviroCon employees.

Before the young man could step into the Jaguar, Jason halted him. "Have many people arrived?"

"Not many, sir; about twenty-five. We haven't had to start parking along the street yet."

"Thank you." The young man kicked up some of the pea gravel as he accelerated away from the esplanade. "He drives like you, Sam."

Sam laughed. "I doubt it. Don't worry. It's my car. If it can survive me, a couple of adventuresome valets won't hurt it."

"By the way, he told me that about twenty-five people have already arrived. Jack said this would be a small party."

"Jason, you still haven't gotten used to the way things work in Washington; have you? By small, Jack probably meant several hundred. If he'd had more time, I'm sure the number would have reached five hundred."

Jason halted. "Several hundred?"

Sam patted his arm. "Yes dear. That's not really very many. Just the EnviroCon people alone will total about a hundred. Besides, I ran into several staffers and Congressmen on the Hill that congratulated me on the engagement and said they looked forward to seeing me tonight."

"But why would they come, especially on such short notice?"

"For one thing, dear, my father used to be a Senator. For another, I am the chief of staff for a powerful Congressman. They know that Geoff will be here tonight. He's begun to test the waters for a run at the Presidency. People will come to talk with him. Besides, EnviroCon is rather well-known for its parties."

Before Jason could say anything more, Mr. Lane's butler,

Gene, outfitted in full livery, opened the door for them. He took Sam's light cloak and Jason's overcoat. "Good evening, Ms. Whitlock, Mr. Graham. Mr. Lane and Mr. Watson are waiting for you in the library."

Jason thanked Gene and escorted Sam to the library. The corridors were brightly lit and hung with the first of the Christmas decorations. When they entered the room, it didn't appear crowded. Jason spotted Charles Lane with an attractive brunette in tow standing next to Jack and Michelle Watson. They walked over to greet them. Jack noticed them first.

"Sam, Jason, come over and join us." Jack was smiling and holding a cocktail in one hand. They walked over to join.

"Congratulations, Ms. Whitlock, Jason. I'm very glad to hear that you two are getting married. I wish both of you the best." Mr. Lane paused to take a sip from his cocktail; a scotch on the rocks from the smell of it. "Let me introduce you to Nicole Wilson." Ms. Wilson extended her left hand toward Jason, her right hand being preoccupied with a martini. He took it for a moment and then released it.

"A pleasure to meet you, Ms. Wilson." She smiled in response, but never really looked at Jason. Rather she seemed to be assessing Sam. "And this is Samantha Whitlock, Nicole."

"I know that, Charlie. Samantha and I have met before."

There was a brief pause, but Sam did not let it last. "That's right. It's good to see you again, Nicole."

"I'm sure." Again, a pause. Everybody but Mr. Lane noticed the tension.

"Nicole and I met when our fathers were in the Senate together. Isn't that right?"

"That's right. We used to run into each other quite frequently. As I recall, you were always trailing along at your mother's heels." Jason came to Sam's rescue. "I'm a bit thirsty myself. Would you like a drink, Sam?"

"I'd love one."

"If you'll excuse us for a minute." Jason and Sam began to walk away.

Jack turned to follow them. "If you don't mind, Jason, I'll join you. I need to freshen up my drink. Michelle, would you like anything?"

"Sure. I'll have the usual."

Jack, Jason and Sam all walked over toward the bar, which had been set up by the cathedral windows at the south end of the library. After they had gotten out of earshot, Jack leaned over and whispered: "What was that all about?"

Jason shrugged. "Beats me."

"It really isn't much. Her father and my father were vigorous opponents in the Senate during my father's term. They rarely agreed on anything and went head to head on a number of issues. I was rather young at the time. Nicole was in her twenties. Her father spoiled her rotten. She was used to getting her own way." Sam paused to order a drink from the bartender. Jason and Jack did the same. "Well, to make a long story short, a few years after my mother died, Nicole started running into my father frequently at social events. Although my father didn't tell me about the episode, a friend of mine just happened to be at the same party the night Nicole got rather drunk and came on to my father a bit too strongly. From what I heard, he rebuffed her in public. To say the least, she didn't take kindly to his public rebuke. It's one thing to be spurned in private but yet another thing in public, especially for a woman like Nicole."

"The few times I've seen her since then, she's either avoided me entirely or made a point to be rude." The bartender handed Sam her drink and then handed a drink to Jason and Jack. "Don't worry, I'll make a point of avoiding her tonight." Sam leaned closer to Jack to avoid being overheard. "But, Jack, I'd warn Mr. Lane about her. She's a real bitch. I heard that she burned through her trust fund and is looking for a sugar daddy."

"I appreciate the warning, Sam. I'll feel him out about her and may take your advice. But he's sensitive about his personal life. I'll have to tread very lightly. Anyway, Nicole is behind us. This is your party. Relax and enjoy yourselves."

Jack returned to his wife and Mr. Lane while Jason used the

convenient excuse of seeing Natalie enter the library to head in a different direction. "Natalie, how are you?"

"Just fine, Jason; fine." Jason could tell by her expression that she didn't want to talk about her mother's condition, so he did not raise the subject.

"Have you met my fiancée, Samantha Whitlock?"

Natalie extended her hand toward Sam. "No, we haven't. It's a pleasure to finally meet you, Samantha."

"The pleasure is mine, Natalie. From what Jason tells me, he'd be lost without you."

Natalie laughed. "You know that isn't true, but I appreciate it anyway." Michael Barnwell appeared at the door and walked up to them very quietly. He was just about up to them when Jason noticed him.

"Hello, Michael. How are you?"

"Just great, Jason." He stopped right next to Natalie. Looking to Sam, he said: "This must be Samantha." Michael extended his hand towards Sam. She took it. "You are the lucky one, Jason. Samantha is both attractive and, from what I've heard, very intelligent." Sam blushed very slightly.

"A pleasure to meet you, Michael. But these compliments have to stop."

"I doubt they will, Samantha. After all, this is a party for you. People would be rude not to make them, even if they weren't true. And they clearly are true." Michael smiled.

"Why, thank you, Michael. I didn't think there were men like you anymore, at least not in Washington." Everybody chuckled. "Treating women like women is not politically correct." Michael turned to Natalie.

"You're looking rather ravishing yourself tonight, Natalie. Let me buy you a drink."

"With a lead-in like that, how can I say no? I'll see you around, Jason, Samantha." When they were halfway to the bar, Sam noticed Michael lean over and whisper something in Natalie's ear. Natalie leaned over rather closer than was needed and their shoulders brushed together as Natalie laughed in response.

"Jason, you didn't tell me that Michael Barnwell and Natalie were an item."

Jason raised his eyebrow. "That's because I didn't know."

Sam laughed. "That's one of the reasons why I love you. Despite your brilliance, you still need me."

Jason and Sam wandered about the party making small talk. This was the first opportunity Jason had to meet the spouses of many of his co-workers. As Sam had guessed, most people began arriving around 8:00. Congressman Barnes and his wife, Virginia, arrived at 8:15. It took them twenty minutes to make their way over to where Jason and Sam had just begun talking with Congressman Ben Slater and his girlfriend, Leslie Jones.

"Congratulations, both of you." Geoff leaned over and gave Sam a hug and a kiss on the cheek and vigorously shook Jason's hand. Virginia leaned over and hugged and kissed Sam as well.

"Thank you, Geoff."

"By the way, Jason, I don't believe that you've met my wife Virginia." Jason extended his hand toward Mrs. Barnes and she took it.

"A pleasure to meet you, Mrs. Barnes. Sam has told me a lot about you." Virginia Barnes smiled.

"My pleasure, Jason. And please, call me Virginia." Congressman Barnes then extended his hand to greet Ben Slater.

"Ben, a pleasure to see you here as well. I enjoy seeing colleagues, away from the Hill."

"You know I wouldn't miss coming to see the man that got Samantha. By the way, this is..."

"Yes, yes, I know, Leslie Jones. You work for the EPA; don't you?"

"That's right."

"I remember meeting you at a committee hearing about a year ago."

"You've got a good memory, Congressman."

Congressman Barnes then took the opportunity to introduce his wife to Ms. Jones. "Now that the introductions are over, when is the wedding?"

"We've decided on March 10. I hope all of you can make it."

"Oh, I'm sure we will." They talked about the wedding for a few minutes and then Congressman Barnes and his wife excused themselves to talk with some other people. Congressman Slater began to excuse himself as well, taking his cue from his mentor to mingle like a politician, but Leslie Jones halted him.

"Jason, there is one thing that I wanted to ask you."

"Oh? What's that?" Jason was unsure of what to expect in light of her tone of voice.

"I understand that you did work on the El Niño phenomena out at Cal Tech."

"That's right."

"Well, I read in the paper this morning that strong El Niño currents have begun to assert themselves for the fifth year in a row. The article I read seemed to attribute the occurrence of five El Niños in a row to global warming. What do you think about that?"

"I'd like to hear the answer to that one myself." Jason turned towards the voice behind him.

Ben Slater scrambled to get around Jason. "Joel, I didn't see you. Jason, let me introduce you and Sam to Congressman Joel Friedman."

Congressman Friedman waived his hand dismissively at the freshman Congressman from New Jersey. "No introduction needed Ben. I've met Samantha and know who Jason is. I'm sure that he knows me. Go on Jason, answer Ms..."

"Jones." Ben Slater interjected.

"Ms. Jones' question. What do you think about all these El Niños?"

"Well," Jason paused. He hoped that his face hadn't reflected his initial startlement at the question or at Congressman Friedman's interruption. He couldn't tell the truth, but it wasn't in his nature to lie either. "To be honest, Ms. Jones, I haven't really thought about the issue. Besides, what I did was design a computer program to analyze and predict the occurrence of the phenomena. I'm not an environmental engineer."

"But don't you have some opinion on the cause of so many El Niños in a row?"

"I seem to recall that just a few decades ago, in the early '80s, we had four, if not five El Niño events in a row. People speculated about global warming back then, but nothing ever came of it. In fact, when I finished my work on my program and reviewed the results of the first few runs, my program predicted last year's El Niño as well as this year's. I doubt we'll have a repeat next year."

"Why not?"

"I'll tell you why not. Because El Niño and global warming is just a bunch of crap. That's why. I've heard enough. Samantha, Jason, congratulations." Congressman Friedman nodded to Sam and then gripped Jason's hand firmly and held it, staring directly into Jason's eyes. Jason could feel the intensity of the Congressman's glare. He wondered about the point behind the Congressman's interruption. Friedman had headed up the double-black Starcom Project that had built a base on the Moon, but he didn't know whether Friedman knew about Project Runaway. From the way Friedman was sizing him up, Jason suspected that he did. Jason tried to release his hand, but the Congressman held it for a moment longer as a smile developed across his face. He then turned quickly and left.

Jason turned back to Leslie Jones. "You asked why El Niño won't repeat this year, let me try to explain. The El Niño event relates to the flow of warm tropical waters across the Pacific Ocean. What we call an El Niño is the condition that exists when the prevailing winds over the Pacific weaken, enabling warmer than normal waters to develop over sections of the eastern Pacific. This in turn creates a wind pattern that blows warm surface waters to the east, against the coast of South America. The warm waters act as a blanket to prevent the usual upwelling of cold waters along South America. The process is rather complex, but, basically, a positive feedback is created that strengthens and perpetuates the El Niño for several years until the system breaks down. Then the reverse effect, called 'La Niña,' replaces the phenomena until a balance is reached again."

"These patterns have been repeating themselves for centuries. There is nothing new about them. It's just that they've only been identified in the past few decades."

"I didn't follow you entirely, but I think I understand. You think these El Niños will end then?"

"If not next year, then they will the following year."

"How do you think that will affect snowfall over the Rockies? I'm originally from Salt Lake and we've had several years of record snowfall."

Jason laughed. "I'm not about to try to predict local weather patterns, but if it snowed a lot during the past few years, it will probably snow a lot this year. However, when the El Niño ends, it's a good bet that the snowfall patterns will change. I'd get your skiing in this year, Ms. Jones."

"Thank you, Mr. Graham. I appreciate it. And good luck with your wedding."

After they left, Jason breathed a sign of relief. "I think I need another drink."

Sam gave Jason a kiss on the cheek. "You handled it well. Friedman's a jerk, a very dangerous jerk. Geoff has crossed swords with him before. Now, let's forget about it and get back to the party."

CHAPTER 10

Toxic fertilizers, sewage and industrial chemicals poured into the South Atlantic from Rio de Janeiro to Sao Paulo as unusually heavy rains swelled the region's rivers. International observers have estimated that the pollution pumped into the Atlantic will create a "dead zone" of over 40,000 square miles along the coast of Brazil and Argentina. Massive oxygen-depleting algae blooms have already begun to appear along the coast and will expand for months to come. Brazilian officials have "temporarily" closed the beaches in Rio de Janeiro, causing a great deal of concern within the Brazilian tourist industry. This summer's extremely high temperatures have exacerbated the effect of the pollution.

⚜

McLEAN, VIRGINIA.

The doctors informed Natalie that her mother would survive into the new year, but barely. While Natalie had decided to continue to work rather than linger at the hospital, she visited her mother every night. The last time her mother regained consciousness was on a tearful Christmas Eve. Her mother, remarkably lucid, had taken the opportunity to say goodbye to both Natalie and to her father.

While Natalie knew she would have trouble accepting the loss of her mother, she was glad that her mother would never find out about her dealings with Kreschenko. Natalie's hatred of Kreschenko grew as rapidly as the cancer gnawing away at her mother. She was dealing with the devil, but she had no choice. Kreschenko had not contacted her during the week following their initial meeting, but he had made his presence known. She had seen him several times outside the hospital — smiling.

Natalie also received anonymous clips from several newspaper articles written by Trey Wilson about Gennady Vorshkoff, including the final obituary that Wilson had written after Vorshkoff had killed himself. But the ultimate insult came when Kreschenko sent flowers to Natalie at her office expressing sympathy for her mother's condition. Natalie had almost broken down and told Michael Barnwell what was going on when he had come into her office to find her sobbing over the flowers, but she maintained her resolve not to bring Michael into it. Kreschenko was already attempting to destroy the lives of the other two people she loved. She had begun to distance herself from Michael in an attempt to protect him, a measure that Michael attributed to her grief over her mother. Although Natalie had decided to deal with Kreschenko on her own, she didn't know what she would do.

To gain time, Natalie informed Kreschenko that she would cooperate with him. Michael had revealed to her in the intimacy of her bed several weeks before of several of the security measures that had been taken. She knew that he had hired several agents to follow various members of the team and that telephone calls were being monitored. As a result, she insisted on meeting Kreschenko today, knowing that the attention of all persons involved would be on the meeting with the Vice-President. As would be expected of her, Natalie would go to her mother's hospital right after Jason and Jack left for the meeting.

After briefly sitting with her father and mother, she would meet Kreschenko in a motel room across the street from the hospital. There she would spell out Project Runaway to him and get his "pledge" not to reveal the information to the wrong party. While she wouldn't put any value in the pledge, she surmised that Kreschenko would not reveal the information because of his secretive nature. She doubted that the type of person that would hire Kreschenko would make the information public. He would use it to his advantage, but he wouldn't make it public. At least that is what Natalie hoped.

She had managed to make an extra copy of the Project Runaway Report while preparing the packages for Jason and Jack. Michael had given her the code to the special copy machine. While she knew that Michael routinely checked the number of copies run, she also knew that he would believe her when she said that the machine had jammed and she had been forced to make an extra copy. Just to be on the safe side, she took out extra sheets of the specially treated paper, ran them through her printer with random text on them and then shredded them in Michael's shredder. She also signed for the paper, signed the shredder sheet and called the copier maintenance service to cover her tracks.

She would deliver the Report to Kreschenko at their meeting. He would, of course, continue to blackmail her for as long as her father lived. Natalie knew that. But she hoped that she would find a way to eliminate Kreschenko before things went too far. Until then she would continue to trickle information to him. In fact, one part of Natalie's mind was relieved at her revealing the information to Kreschenko. While she agreed that the information must be kept confidential, she also thought that other nations should not be left out of the loop entirely. And although she didn't buy Kreschenko's line about helping Mother Russia, she did hope that those in Russia who ultimately obtained the information would make good use of it.

❖

Jason reviewed Jeff Nakamura's progress report on the disinformation program that the Vice-President had directed them to undertake. Things were going smoothly. Jeff had almost finished falsifying the data. Jason knew that the truth would eventually come out. After all, in ten years it would be very difficult to hide. But the dummy data, together with "spending cuts" that would curtail additional research, should buy them the time needed to implement Project Runaway.

He had done a bit of unofficial research on the Vice-President over the past few weeks. The results impressed him. The Vice-President had graduated cum laude from Dartmouth College with a BA in international studies and matriculated at The University of Virginia School of Law. There he served as an articles editor of the International Law Journal and graduated Order of the Coif. Upon graduation he had clerked for two years with Judge Rheinhold of the 11th Circuit Court of Appeals and subsequently took a job as Assistant District Attorney for the City of Charlotte, North Carolina. After a brief but stellar career as a district attorney, he successfully ran for Congress, where he served for five terms before being elected Vice-President.

All of the information that Jason read about the Vice-President indicated that he kept his word and was known for his integrity. Few blemishes appeared on his record. Jason took comfort from the fact that Vice-President Hammond, rather than President Wright, was heading up the Program. Jason did not know what to expect at his meeting with the Vice-President, but he looked forward to it.

Jack Watson walked into Jason's office at precisely 9:30. "Are you ready to go, Jason?"

"Almost; I'm just waiting for Natalie to bring up the additional copies of the revised Report. Speak of the devil..." Natalie had just entered Jason's office.

"Good morning. Here are the copies you'd asked for, as well as your original. I'm sorry it took so long, but I had a problem with the copy machine. One of the copies got screwed up and I had to shred it. But I checked these. They're ok."

"Good. Thanks for coming in today, Natalie. I appreciate it."
Natalie turned to leave.

"Oh, Natalie?"

"Yes, Jack."

"I know it's difficult. But please try to have a Happy New Year."
Jason nodded in agreement. "The same goes for me, too. I
know that you've heard it before, but if you need anything,
don't hesitate to call."

"Thank you both. And Happy New Year." Natalie turned quickly before either of them saw her tears. She hoped they would, if
not forgive her, understand what she was about to do.

Jack and Jason drove out the Lee Highway towards the Blue
Ridge Mountains to where they were to meet with the Vice-
President and the other members of Project Runaway. Michael
Barnwell had gone out the night before to meet with the Vice-
President's security detail and work out security issues. Sam was
not going with Congressman Barnes, but Jack had agreed to drop
Jason off at Sam's father's house on the way back from the meet-
ing. The house was only twenty minutes off the Lee Highway.

They left the highway shortly after entering the Blue Ridge
Mountains and took a number of small country roads. Jason
began to wonder whether they'd taken a wrong turn when they
arrived at a rusty iron gate attached to a moss and vine covered
stone wall. Jason couldn't see farther than a few yards from the
road due to the height of the undergrowth. "Are you sure we're
in the right place? It doesn't look as if anyone has driven through
this gate in months." Jack only smiled.

They waited as the gate slowly opened. Jason was surprised it
could open at all, let alone electronically. He looked back after
Jack drove the car in through the gate and could just discern the
mechanism that had opened the gate under two hedges. It would
not have been visible from the other side of the gate. They
rounded a sharp bend in the road and the scenery changed dramatically. The twisted hardwoods choked by virtually impassible
undergrowth abruptly ended, to be replaced by a well-mani-
cured lawn spreading out approximately forty feet on either side

of the driveway, which itself had been transformed from a rut-
ted dirt country road to the pebbled drive of an upscale country
estate. A grove of apple trees extended off down a hill to the left
and a tall hedge obstructed Jason's view to the right.

The road split as the hedgerow ended. Jack made a right
turn. To Jason's surprise, a modern one-story office building
and parking lot that looked as if it belonged in an Alexandria
office park rather than in the Blue Ridge Mountains emerged
from behind the hedgerow.

Jack noticed Jason's surprise and smiled. "Impressive, isn't
it? Geoff Barnes warned me about what to expect. FEMA built
this place back in the 1980s as a retreat for key governmental
personnel in the event of war. The gate and the road are cam-
ouflaged to keep out unwelcome visitors."

Jack pulled the car around to the back of the building. Jason
could see Marine 2, the Vice-President's helicopter, resting next
to the tennis court. After parking they both got out and walked
over to the entrance to the building. They walked through the
first of two sets of automatic glass doors. Jack and Jason both
almost walked into the next set of glass doors because they did
not open immediately. Before they could think of what to do,
the doors opened and a Marine corporal met them.

"Good morning, gentlemen." He handed each of them a
small pin. "Please pin these on an article of clothing that you do
not plan to take off. They will serve as your identification
badges. Now, if you will follow Private Fisher, he will escort you
to the meeting room."

Private Fisher entered from a room behind the corporal's
desk. They thanked the corporal and followed the private down
a hall to an elevator. Jason knew that there wasn't a second floor
and assumed that they would be going to a basement. However,
when he entered the elevator, he noticed seven buttons. The private pushed the first button.

"Excuse me, Private. Are there seven sublevels in this building?"

The private did not change his stance or his expression. "I'm
sorry, sir. I'm not at liberty to discuss the layout of this facility."

The doors opened before Jason could say anything else. They followed the private down a nondescript, well-illuminated corridor past several doors. This building looked like most other new government offices that Jason had seen; economical and unappealing. They turned down one corridor and Jason saw several men standing outside a door. When they got closer, Jason recognized them as members of the Vice-President's Secret Service team. Jason felt a little uncomfortable as the men eyed him, Jack and, to Jason's surprise, the Marine private.

One of the men whispered something into his collar and opened the door as they approached. They didn't say anything to Jason or to Jack. The private stood to one side and extended his arm, "This way, gentlemen." They thanked the private and entered the room.

The room was larger than Jason anticipated, approximately 30 feet by 50 feet. A large black granite table and fourteen chairs dominated the room. Jason noticed that the Vice-President, Congressman Barnes and several other people were standing at the opposite end of the table, in front of a window in the midst of a discussion. They hadn't noticed Jack and Jason enter the room. Jason followed Jack along the right-hand side of the table. Something felt wrong about the picture he was looking at, but he couldn't quite figure it out. When they had gotten closer to the group standing in front of the window, Jason remembered that they had started on the ground floor and had gone down! He looked more closely at the "window." It showed the parking lot, Jack's car and the Vice-President's helicopter. He realized that it was a closed circuit image of the parking lot, not a window.

Congressman Barnes noticed their approach and turned to meet them. "Gentlemen, nice to see you again. Glad you could make it." The Congressman obviously knew that they couldn't have missed the meeting, but Jason didn't detect any cynicism in his greeting. Evidently, the Congressman was just being polite. Jack, however, had picked up on the irony in the Congressman's greeting and decided to respond. "Miss this, Geoff? We wouldn't miss this meeting for the end of the world."

The Vice-President evidently overheard what Jack had said and turned towards Jack with a chuckle. "I'm glad that somebody still has a sense of humor here; and I am glad you could make it, Jack. You, too, Jason." Jason and Jack both simultaneously thanked the Vice-President.

"Now, why don't we begin? Everybody take a seat and I'll make the introductions while we sit down." The only ones Jason recognized were Jack, the Vice-President, and Congressman Barnes. Jason did not recognize either of the other two men, one dressed in an Air Force general's uniform, the other in a business suit, or the one woman, dressed in a grey suit and cream colored blouse. They took seats to the right of the Vice-President, the general sitting down the fastest in order to secure the seat closest to the Vice-President. The other man sat down next and then the woman. Congressman Barnes sat to the left of the Vice-President with Jack next to him. Jason took the seat next to Jack.

"I'll start on my right. This is Frances Crawford. She is the Assistant to the National Science Advisor in environmental matters. Gene Samuelson has elected to take a less than active role in Project Runaway and suggested that Ms. Crawford act in his stead."

"Next to Ms. Crawford is Hamilton Smith. Mr. Smith is with the CIA and will be in charge of security for the Project. It is my understanding that you have already met with Michael Barnwell about security?" Mr. Smith nodded.

"On my right is General George Maxwell. General Maxwell headed up the day-to-day operations of the project that developed the Lunar Base. Immediately on my right is Congressman Geoff Barnes whom I'm sure all of you recognize. To his left is Jack Watson, one of the chief architects of EnviroCon. And finally, to Mr. Watson's left, is Jason Graham, the computer genius that devised the programs that led to this meeting."

"I called this meeting for several reasons, the first of which is to organize Project Runaway. But before we get started, I'd like to turn the floor over to Ms. Crawford. Ms. Crawford. . ." The Vice-President took his seat.

Ms. Crawford remained in her seat. She pushed her glasses back up on the bridge of her nose, pulled some papers out of her briefcase and arranged them on the table. Jason noticed that she held a light pen in her left hand. "Gentlemen, Gene Samuelson approached me immediately after his initial meeting with Mr. Watson and Mr. Graham. He indicated that you had convinced him that we were in the midst of a major environmental catastrophe and wanted my input. Initially, I didn't believe him. While I have studied the damage that we have inflicted on this planet throughout my entire adult life, I found it hard to believe that we could actually destroy our ecosystem."

Ms. Crawford paused briefly. "However, Gene insisted that I review the Report. I did. Your work overcame some of my initial skepticism, but not all of it. I wanted independent verification. I said as much to Gene. I met with him and the Vice-President and they directed me to conduct a thorough check of your work. I completed my report last week. Gentlemen," she paused again, "unfortunately for all of us, I find no fault with your programs or the data. In fact, in certain areas I found that you were overly conservative. We face a global catastrophe. I'm here to work with you and to lend you whatever support I can. Both official and unofficial."

The Vice-President responded. "Thank you for your summary, Ms. Crawford." He leaned back in his chair and put his hands behind his head. "Well, Ms. Crawford, gentlemen, shall we get started?" The first thing they did was to witness the execution of the implementation document by the Vice-President, Congressman Barnes and General Maxwell. They were given a copy of the document, rather shorter than Jason would have anticipated. It recited the Executive Order that enabled double-black projects. The Order granted the authority to take "all actions necessary" to keep the project secret and protect the best interests of the United States of America.

Next, Vice-President Hammond and Congressman Barnes elected the individuals present in the room as the members of the executive committee of Project Runaway. General Maxwell gave everybody a briefing of the capabilities of Starcom, the lunar base that the Air Force had previously constructed with double-black funding in connection with the Strategic Defense Initiative.

General Maxwell also briefed everybody on the layout of the complex they were currently in. Built by FEMA, the Federal Emergency Management Agency, during the early 1980s, the facility had served as a top-secret retreat for officials in Washington in the event of a nuclear war. After the breakup of the Soviet Union and the end of the Cold War and as a result of significant bad press that FEMA had received after Hurricanes Andrew and Hugo, their focus had changed from preparing to survive a major war to responding to natural disasters. Starcom had purchased the facility from FEMA in the early '90s and had retrofitted it for their purposes. The facility remained well hidden from public scrutiny, but was close enough to Washington to serve as the headquarters for Project Runaway. It had its own small nuclear generator and a state of the art CRAY III computer.

Jason cleared his throat at the mention of the Cray III and it being state of the art. The Vice-President, evidently noting Jason's point, interrupted to let everybody know that they intended to obtain an Omegatron in the next year; but, until then, they would continue to link into NASA's Omegatron.

General Maxwell, not used to being interrupted, gave Jason a brief glare, but nothing else. He knew better than to reproach the Vice-President. The General finished his report on the status of the facility. It sounded ideal for the Project. While the complex was far larger than necessary for their current purposes, eventually, the Project would expand and fill the complex.

Hamilton Smith gave everyone a brief rundown of the security measures that would be taken in connection with the Project. Jason got the distinct impression that he would not want to get on Mr. Smith's bad side. Jason always felt a bit ill at ease around Michael Barnwell, but Hamilton Smith made him squirm in his seat. Jason noticed that everyone, excepting the two politicians, looked a bit uncomfortable. While Jason had initially questioned the security for the complex, he had no doubts about it now. The undergrowth and broken down fence that he and Jack had seen actually served as a cover for state of the art surveillance equipment and an "active" defensive perimeter that Mr. Smith merely alluded to. Everything having to do with Project Runaway would be on a need-to-know basis, even for the executive committee.

The Marine guards at the complex would not know the purpose of the complex and were trained not to question the nature of their assignment. Their commander, a major, had been told that the base remained a top-secret retreat for Washington brass in the event the "balloon" went up. He did not question the fact and enjoyed the assignment, given the proximity to Washington, D.C.

Hamilton also informed them to report any suspicious activities and to never use the telephone to discuss the Project, even secured lines. With something this sensitive, they could take no chances.

Congressman Barnes next took the lead to discuss the logistics of the Project. Given his visibility as well as that of the Vice-President, they couldn't take an active role in the Project. However, they would get periodic updates and would need to be a part of any major decisions that were to be made. The Congressman indicated that given Samantha's impending marriage to Jason, as well as her work with the Congressman, she would make the perfect intermediary to keep the Congressman and the Vice-President up to date. She would not raise suspicions, whereas too many meetings between EnviroCon personnel and either the Congressman or the Vice-President would result in too many questions.

The Vice-President would see to General Maxwell's reassignment to Vandenberg where he could head up the space operations necessary for the Project. The General would run the testing of the initial space elements necessary for the system as well as coordinate the Air Force's role in the soon-to-be expanded Space Station Freedom. Fortunately, as a result of the actions of the Chinese, the Congressman did not doubt that they would be able to dramatically increase funding for space once again. With the support of the Vice-President, he felt virtually certain that they would get the budgets necessary to move Project Runaway forward for a couple of years.

As a member of the administration, Frances Crawford would need to remain in Washington. She could not dedicate all of her time to the Project; however, over time, she could delegate more and more of her work to others to enable her to increase her focus on Project Runaway. She had also indicated that she would resign her office in the next year to work full-time on the Project. Jason looked over at her. It suddenly dawned on him how seriously they were taking the Project. A young and rising political star would give up or, at a minimum, seriously hurt her political future to dedicate her time and effort to the Project. While Jason did not doubt that he would do the same, his respect for her increased.

As a CIA agent at large, Hamilton Smith had the discretion to assign himself to Project Runaway and would work out of the Blue Ridge facility. However, given the nature of his duties, he would be out of pocket frequently.

Finally, the Congressman turned to the roles of Jack and Jason. Because of Jack's long-time relationship with EnviroCon, he would remain with the firm and would oversee a number of studies that would be necessary in connection with the Project. Charles Lane was on the verge of retirement — Jason hadn't known that — and Jack would take over the reins of the firm.

The Congressman then turned to Jason; "Jason, we need somebody on the executive committee to oversee the general operations and planning of the Project. You've heard what I said about the roles of all of the other members of the executive committee. For a variety of reasons, most of which relate to our public visibility, none of us can take on the task of running the Project. But you can. You are new to Washington and are a virtual unknown. You can leave EnviroCon and will not be missed. There will be some scandal, but Jack can control it. Do you understand what I am getting at?"

"I think so, Congressman." Jason's thoughts of a few minutes before about the sacrifice of Frances Crawford had come home to roost. "I'll resign whenever you feel the time is right, immediately if necessary."

"Unfortunately, Jason, that's not the way we need to work things. I wish it was."

Jack looked Jason directly in the eye. "We've got to discredit you." Jason felt all the eyes in the room focus in on him.

"I'm not sure that I follow you, Jack. Why?"

"Look, Jason. Think about it. It's your programs that led to the discovery of the runaway greenhouse effect. Without them, the information would not have come to light for years."

"I'm not sure that's true, Jack. Given the capacity of the Omegatron, somebody would have devised a system to correlate the data in the next few years."

Frances Crawford spoke up. "I've got to disagree with you, Jason. You've underestimated your own capabilities. I used to be a fair programmer, but I couldn't follow the steps that you took. I had to get your former professor, Walter Bennette, to help me out. Your program works, but Bennette and several others that I checked with doubt that anyone will replicate it for at least a decade. That will give us the time we need."

"The problem is, Jason, your El Niño thesis is on record. By using it, other computer whizzes will shave years off of that decade; and we need all ten years. We've got to discredit you, and publicly, so that others don't look to your research. We actually ran the hypothetical through your own programs. They confirmed our thoughts. If we discredit you, the steps you took will not be repeated for years. Otherwise, your own programs predict that somebody will duplicate your work in a few years. Do you think that the rearrangement of the data would hold up to the scrutiny of your programs?"

"They'd pick up an anomaly." Francis Crawford nodded in agreement.

"EnviroCon will fire you for your incompetence. That will buy us several more months to "fix" your screw-ups while we complete the falsification of the NASA data banks. But that's not all. I spoke with Walter Bennette personally and briefed him on as much of the situation as I felt I could. After a great deal of arguing — he's a big supporter of yours — he agreed to discredit your graduate work as well."

"What do you mean, discredit my graduate work?"

"Professor Bennette is going to have your Masters Degree revoked. He'll claim that, upon further review of your thesis, he discovered that you plagiarized key sections of it and falsified your data."

Jason was too shocked to respond. He didn't mind quitting his job or getting fired, but to discredit him, that was far more than he had expected. He'd never be able to get another job working on a computer except with the Project. They would have him by the balls. As much as he knew, intellectually, of the necessity of the action, he still couldn't quite accept it. "I hope that you all see the irony of discrediting me to enable you to complete precisely the type of work that you will be accusing me of having undertaken!"

"We know, Jason. If we could avoid taking the step, we would."

The Vice-President spoke up. "Jason, we have no choice. All of our futures are under a death sentence that you delivered. It's now up to all of us to do everything within our power, at whatever the cost, to commute that sentence."

"I know you're right and I agree with the decision, but that doesn't make it any easier."

"Jason, we'll delay the action until after your wedding, but then we'll act. We need you up here to head up the Project."

"Congressman?"

"Yes, Jason."

"Does Sam know about this?"

"No, Jason. I thought that you'd want to discuss it with her. But I know Sam almost as well as you do. She'll understand. Just as you do yourself."

The meeting went on, but Jason had trouble concentrating. The initial steps to be taken under the program were set. The

Vice-President and Congressman Barnes would work to secure additional funding for the Space Station Freedom to enable a rapid expansion of a manned presence in space. One of the features of the new funding would be extensive testing of construction techniques to enable the construction of larger structures, ostensibly to "compete" with the new Chinese presence in space. In addition Congressman Barnes felt that he could get an appropriation through to include the funding of a small solar powered satellite and receiving antenna, both of which would be critical during later stages of the development of a space city.

General Maxwell, with the help of Congressman Barnes and the Vice-President, would shift funding from other double-black projects to Project Runaway. He could control the testing of new launch and orbital transfer vehicles from Vandenberg. While they didn't need to go back to the Moon yet, the General would keep the base on line. He would also restart the testing of a mass driver that the Department of Defense had investigated as a potential ABM system.

Jack and Jason would finish the disinformation program and Jason would gear up to take over the overall management of the Project. They would create a dummy corporation to obtain the direct contracts with the government and would subcontract out elements for actual research and development.

Finally, before the meeting broke up, Jack nudged Jason to remind him of the one new element of the Project that they had brought to present to the group. Jason cleared his throat to talk.

"Ms. Crawford, Gentlemen, I appreciate your confidence. You have given me a great deal to think about today and I'm not prepared to review all of the steps that I feel will need to be taken with the Project. However, while I'm grateful for the confidence you have shown in me, I need to point out that I have only minimal experience as an administrator. We can't afford to screw up. I'm willing to serve in the role you have chosen for me. However, I'm not qualified to be an administrator."

Congressman Barnes interrupted. "We know that, Jason. We are providing you with the best support we can in the form of several administrative assistants. In fact, you'll be meeting one of them, a Major Ryan, shortly. He'll be in to meet with you and Jack after our meeting. Don't worry about the administrative end, Jason. We've taken care of you."

"Thank you, Congressman. It gives me more confidence to know that you are intelligent enough to see my shortcomings. Now, where was I? Oh yes, one of my first tasks will be to prepare a chart outlining the general steps that we will need to take. We can leave that until our next meeting. However, as a stopgap, I've prepared a critical time line. The work is preliminary, but it will give everybody something to work on. I've attached it to the front of my Report." Jason pulled copies of the Report out of his briefcase and handed them out around the table.

"If each of you could respond to the projections I've prepared, I would appreciate it." Everybody looked over the two pages of the time line. "I don't expect you to respond now. Take your time and get them to me later."

Hamilton Smith had a bit of a frown on his face. "One comment, Jason — while you printed these Reports on copyguard paper, there are a few things that you should change. First, we'll know where these Reports are coming from because they will all need to be delivered by hand. Remove all references to EnviroCon, Project Runaway and all names from all future Reports. That way, if a copy does fall into the wrong hands, they won't know where the information came from. We all know the basics of the Report and of the problem we face. In all future communications, break information out into separate reports. That way, if a report falls into the wrong hands, it won't reveal the whole picture. This Report is good, too good. If we lost a copy, it would give away too much information. I've already reviewed security with Michael Barnwell to make sure that these and some other steps are taken."

Jason could only respond with a half-hearted grin. "Thank you, Hamilton. That just goes to emphasize my point that we all need to give our input on this Project. While we do have some overlapping experiences, we each have our own special areas of expertise to contribute."

The Vice-President looked down at his watch. "I've got to go. I think we've accomplished a great deal with this initial meeting. Jason, I'll give this my highest priority and will keep in touch with you. Now, if you'll excuse me." Everybody stood up as the Vice-President left the room.

Congressman Barnes took over where the Vice-President had left off. "This timetable looks pretty good, Jason. I think that you are being optimistic about when the public funding comes on line, but you aren't off by very much."

"I'm glad of that, Congressman. According to my calculations, we don't have a lot of room to play around. Start up is crucial. If we don't meet the early dates, we're bound to miss our later targets. We can't afford to do that."

"I appreciate your candor. The Vice-President and I will do all that we can to move the public aspects of the Project forward. Unfortunately, we are dealing with politics. We will be as convincing as possible, but we're bound to run into unexpected hurdles. If not for the Chinese, we'd never get an increase in appropriations. Fortunately, with Mr. Smith's help, we'll tie public suspicion of Japanese involvement with the Chinese to create a new "Red" scare; but, even so, it won't be easy."

The General jumped in on the point. "You're right there, Congressman. We know that the Chinese are capable of the launch on their own, but the public doesn't know that. Our intelligence predicted a launch last year. The Chinese have been itching to show off. I met General Liu, the head of their space forces, at a Pacific Rim conference a few years back. He'll keep pushing, which is just fine for our purposes."

Frances Crawford leaned forward and put her pencil down. "I know several members of the press. I could leak information implying that the Japanese helped the Chinese out with their space program. They'll name a person 'close to the President' as their source. The Administration will, of course, deny the information, but the press should dive right into the fray. Given the amount of recent investment by the Japanese in China and our trade problems with Japan, it will not be difficult to swell public sentiment against the Chinese."

"That's a good idea, Frances, as long as you can trust your contact not to reveal your name. As you know, the Japanese can, and will, bring a great deal of pressure to bear."

"I know, Jack. But I know my source. He won't reveal his sources. By the time the Japanese and the Chinese make it clear that they're not working together, the damage will already be done."

Congressman Barnes checked his watch. "Good Frances. Check with the Vice-President before planting the leak, but I think that it will work. I've got to go as well. Jason, you and Jack are planing to stay for a while; aren't you?"

"Yes, we are. When we're done, Jack is going to drop me off at Sam's father's house for the holiday before going back into town."

"Give Sam any information you need to pass along to me. I'll get it from her after the first. In the meantime, I hope all of you have a Happy New Year!" Frances Crawford also excused herself from the meeting, as did Hamilton Smith. The General then reached for his hat.

"I've got to go as well, gentlemen. I have a flight back to Vandenberg waiting for me at Andrews. But before I go, I'd like to introduce both of you to Jason's new administrative assistant and my liaison." The General pushed a button. A major in a blue Air Force uniform walked into the room. Jason, busy saying a last word to Frances Crawford, didn't notice the major enter. Frances left and Jason turned to meet the major. When he did, he got quite a shock. The major was Bob Ryan, his old roommate from MIT. Jason began to say something but Bob extended his hand towards Jason before he could.

"Nice to meet you, sir." Jason looked at him quizzically, but went along with him.

"My pleasure."

After Major Ryan was introduced to Jack Watson, the General interceded. "You are in capable hands, gentlemen. Major Ryan and I have worked together for several years, including on Starcom. He's a top logistics officer with extensive computer expertise and a top security clearance. He'll show you around the base. Major?" The General turned to Major Ryan. Major Ryan gave him a crisp salute, which the General promptly returned. "Thank you, General."

"Now if that is all, gentlemen, I'll be leaving. I'll expect periodic reports from you, Major."

"Yes sir, General." The Major gave the General a crisp salute.

After the General left, Jason turned to Bob Ryan. "Excuse me, Major. But I've got to go to the rest room. Do you know where one is?"

"Yes, sir; but it's difficult to find. I'll take you there. Mr. Watson?"

"I'm fine. I'll wait here for you."

After they got out in the hall, Jason leaned over to Bob. "Bob, what the..." Bob signaled him to stop. They walked down the hall and into the men's room.

"Jason, before you say anything, let me explain. The bathrooms are the only rooms in the building that aren't monitored. While I doubt that anybody would check the tapes of our conversation in the hall, I'd just as soon not take the chance."

"But why, Bob? What's the big deal?"

"I really want this assignment. I had to work hard kissing Maxwell's ass to get it. I don't want to take a chance blowing it."

"But how would acknowledging that we know each other blow your assignment?"

"It might not. But knowing the General, I think it would. He'd have no real grounds not to choose me, but the choice is his. He doesn't need a reason. And he is a stickler for undivided loyalty. If he knew of our past friendship, he would guess that I might have some loyalty toward you. He wouldn't like that. He wants his troops to be loyal to him and him alone."

"When he brought me in to review the new assignment, it was just what I was looking for, a chance to get out from under his constant scrutiny. He'd been grooming me for a role like this, to serve as his eyes and ears in a project of this importance. If he knew that we used to be roommates, he'd have chosen somebody else."

"But won't he find out that we roomed together? After all, it's in your file as well as mine."

"That's true, but he'd never go to the trouble of checking. Hamilton Smith is the only person that might discover the link. And, from what I've heard about the General's prior run-ins with the CIA, I doubt that Mr. Smith will share the information with the General. After all, it's not the type of information that would harm the Project. It doesn't make me a security risk."

"Well, I'll go along with you, Bob. It sounds like the General is the type of person that I need to be leery of."

Bob nodded. "My position on his staff can only help you. So let's keep our relationship a secret. If it ever comes out, which I doubt, we'll just say that we weren't on speaking terms after we graduated from MIT due to a problem with a woman." Bob smiled. "Given my reputation, that will be believed."

"We do have a lot of catching up to do, Bob. Whatever led you to sign up with the Air Force? I never would have expected it."

"It's a long story. But let's get back before the others get suspicious." Bob began to walk towards the door.

"Oh, one thing that can't wait, Bob." Bob turned to face Jason. "Sam and I are getting married!"

"That's great news, Jason! Congratulations! In fact, I expected you two to get together before we left MIT. But anyway, we've got to get back. Why don't you come back up here in a few days. I'll arrange to give you a tour of the grounds. That way we can catch up on things out of earshot. Give Sam my best. Actually, you'd better not, this Project is too sensitive."

"That's not a problem, Bob. Sam is part of the Project. She's Congressman Barnes' chief of staff."

"Great. Then give her a hug for me."

✦

CHEVY CHASE, MARYLAND.

Natalie arrived at Room 101 of the motel, got the key out of her pocketbook and went in. She could smell the noxious smoke from Kreschenko's cigar the moment she entered the room. But where was he? She heard the toilet flush. He walked in buckling up his pants.

"Ah, Natasha, I didn't hear you come in, but I was . . . preoccupied. Have a seat." Kreschenko pointed to two foldout chairs he had set up in front of a small folding card table. He had thought of everything. Natalie hadn't thought about where they would sit. If he hadn't brought the table, they would have had to sit on the bed. The thought of that made Natalie shiver.

"Thank you." She took her seat. He sat down as well. She withdrew a cigarette from her pocketbook and lit it. It would keep her from smelling the garbage she was sitting next to. He waited for her to say something.

"As I indicated before, I've agreed to cooperate with you. But I'm not doing it for money. What I want, no, demand, in exchange for my cooperation is your original file on my mother and father." Natalie doubted whether even getting the original files and destroying them would do any good because he would have copies. But she also felt that he would expect her to ask for them.

"Natasha, that will be very difficult. But, if the information you give me is worth it, I will try."

"Oh, the information will be worth it. I'll expect the file next time we meet."

"Okay, I'll see what I can do."

"Just do it. Now, before I give you any details on the information, I need to get an absolute assurance from you that this information will never be made public."

After a brief delay, Kreschenko smiled. "Of course I'll keep it confidential. We will make use of the information and destroy it."

"I'm deadly serious about keeping the information confidential. If you don't, the consequences will be dire, not only for you and I, but for Russia and for the United States."

Kreschenko's eyes narrowed. "Look, Natasha. I came over

here to help my country, your parents' country. You might disagree with my methods, but my intentions are good." Natalie
looked him right in the eye. He stared back. She couldn't tell
whether he was telling the truth, but it didn't really matter. She
had no choice. Besides, she hoped that whomever was employ-
ing Kreschenko would have the good sense to keep the information secret.

"Okay." Natalie pulled a copy of the Project Runaway Report
out of her pocketbook. "This is what I'm giving you." She
handed him the Report. He glanced at it, saw the security clearance typed on the cover and smiled. "I suggest that you read it,
especially the section in there that discusses keeping the Report
an utmost secret, before deciding who to give it to."

"I will, Natasha. I will. And thank you."

Natalie got up to leave. "I'll expect the original files next
week. By the way, leave a message at the hospital for me, not at
my office."

"I'm not stupid, Natasha. I'll take care of it."

"Good." Natalie got up and left.

Kreschenko sat back in his chair, relit his cigar and began to
read the Report.

CHAPTER 11

Captain Victor Solokoff of the icebreaker V.I. Raskolnicoff paced unhappily along the bridge. He faced the worst year in his company's history. He and a half-dozen fellow captains of the old Soviet fleet had formed the company several years before and had done very well. Soviet icebreakers had always been regarded as the best in the world. Russia had very few ice free ports. His company made a great deal of money the first few years of its existence, but this year, nothing. The entire White Sea remained free of ice. Traffic flowed easily around the North Cape. Ice had yet to snare traffic in the Gulf of Finland. There was no work. He had hoped for a major front to build up, but the winter remained the mildest in memory.

✦

Moscow, Russia.

Nicholai Stipanovich Greschko stood behind his desk on the fifty-sixth floor of his Moscow office building, the first of the truly western skyscrapers built in Moscow. As one of the de facto rulers of a new, capitalist Russia, he liked the irony of his view, looking down on the Kremlin. For some reason, his lackeys in the Russian Parliament didn't appreciate that irony.

He stood at the window looking down on the cold, dank streets of Moscow. Few people were evident after the revelries of New Year's Eve. Because of communism, New Year's Eve had become the major winter holiday in Russia. Grandfather Frost brought gifts on New Year's Eve rather than Christmas Eve. While many Russians had returned to celebrating Christmas after the fall of communism, more had continued to celebrate New Year's Eve and New Year's Day as Westerners did Christmas. Christmas remained a religious rather than a commercial holiday in Russia.

There was a knock at the door. The door opened. Vladimir Kreschenko walked in. Kreschenko knew better than to sit down until Greschko did. It reminded Kreschenko of the old days in the Lubyanka. Greschko was an old time Russian and Kreschenko knew better than to cross him.

Greschko finally turned. "Happy New Year, Volodia."

"Happy New Year to you, Nicholai Stipanovich."

"You know, my children were not happy about my leaving this morning, Volodia." Greschko took a seat behind his desk. Kreschenko took a seat in the small wooden chair in front of Greschko's desk. Kreschenko didn't know whether the legs had been cut back on his chair, but he was forced to look up at Greschko, over the heavy, ornate desk.

"I'm sure you will not be disappointed, Nicholai Stipanovich." Kreschenko pulled his copy of the Project Runaway Report out of his briefcase, stood up and handed it to Greschko across the desk. "I have read the Report, but I have not had it translated yet. Given the content I did not know who to trust in its translation. You will find my summary at the front of the Report."

Greschko leaned back in his chair and began reviewing the Report, ignoring Kreschenko's presence. Twenty minutes later

Greschko looked up from the Report. Kreschenko smiled, expecting to see some reaction from Greschko, but none was forthcoming. Greschko then turned his chair to stare out his window. Kreschenko squirmed in the silence. Finally, he could not stand the silence any longer.

"Did you understand the Report, Nicholai Stipanovich?"

"Of course I did! I'm thinking. Do not interrupt me again!" Kreschenko waited in silence for several more minutes. When Greschko's chair turned back around, he was smiling. "Very good work, Volodia, very good. But are you certain of its authenticity? There are many in the West that would love to discredit me."

"I'm certain, Nicholai Stipanovich. After reading the Report, I used many favors to check the information. The firm had access to a new computer the Americans have developed and had a contract from NASA to analyze satellite data. They've met several times with both Congressman Barnes and with the Vice-President. Read the end of the Report. It will tell you what I believe they are planning to do."

"I'll get to that soon enough. It is what I will do that interests me now." Greschko pushed a button on his telephone. Immediately a middle-aged man appeared at the door. "Alexander, please go make me several copies of this and call Volkov. We need to get it translated immediately."

The man took the Report from Greschko and left the room without speaking. "Now, Volodia, I want you to keep on this source of yours. If possible, I want copies of everything you can get. See if you can't get computer records. They will be the most helpful."

"Yes, Nicholai Stipanovich. I'll do that immediately." Kreschenko got up to leave.

As he was getting up, Greschko addressed him again. "Oh, and, Volodia, I believe that Grandfather Frost may have just gotten you that dacha in the Urals that you've been wanting."

"Thank you Nicholai Stipanovich." Kreschenko turned to leave as Alexander ran back into the room, his face ashen.

"What is it, Alexander?"

"I'm very sorry, Nicholai Stipanovich. But the copy machine must have malfunctioned."

"What do you mean?" Greschko stood up from behind his desk.

"Look at the paper, Nicholai Stipanovich." Alexander placed the Report on the desk. The paper was black, totally illegible. Greschko looked at the Report. He picked it up and flipped the pages. They were all black. "What do you make of this, Kreschenko?" He thrust the Report at Kreschenko.

"It must be some kind of new paper that can't be copied, Nicholai Stipanovich."

"Goddamn it, Kreschenko! Why didn't you find this out?" Kreschenko had no answer. "Don't just stand there, get me another copy, immediately!"

"Yes, Nicholai Stipanovich." Kreschenko exited the room as quickly as he could without running.

"Now, Alexander, get Volkov up here immediately. The lab should be able to get something off this paper. After all, that's what I pay them for. For your sake, I hope so. GO! Don't just stand there." Alexander left.

Greschko turned his chair after Alexander left to look out the window on the streets of Moscow. Despite the setback with the copying of the Report, Kreschenko had made his day. His agents in Parliament had informed him that funding for space ventures was to be cut back again. They could no longer block the vote. Even though he had managed to diversify into the international civilian aircraft and space launch vehicle market, Russia's military and civilian space programs still constituted a large percentage of his business.

Between the machinations of the Chinese in orbiting a manned space station and the news he had received from Kreschenko, it looked to be a prosperous new year. Regardless of whether the Report was correct, he could use it to bully several key politicians and industrialists into joining his camp. Who would not want to hedge their bets in the face of the Report. He envisioned the lucrative contracts that he would receive in order to construct new launch vehicles to take Russia to Mars and to construct a base on Mars. Let the Chinese have a space station. Let the Americans build colonies at the Earth's Lagrange points. Russia would have the high ground. Russia would have Mars.

And he, Nicholai Stipanovich Greschko, the son of a minor party bureaucrat, he couldn't lose. If the Report proved false, he would make millions, no billions in new contracts. If the Report proved true, he would rule a new planet, away from the impending catastrophe.

CHAPTER 12

A low-pressure cell began to form 900 miles off the coast of Morocco. Unusually warm waters fed moisture into the storm, building its strength. Several ships reported heavy squalls developing and spinning off of the storm. Satellite pictures received in the National Hurricane Center in Coral Gables, Florida, are glanced at and then filed away. If this low-pressure cell had formed several months before or several months later, it would have received the attention of the meteorologist in charge. However, hurricane season was months away.

✦

Jason and Sam slipped into their suite at Casa De Campo, exhausted. The wedding, the reception, changing planes in the Miami International Airport and the drive over bumpy roads from Santa Domingo, had drained them. The bellboy dumped the bags in the room, Jason tipped him and he left. They closed the blinds, crawled into bed and fell asleep in each other's arms.

Jason awoke to a persistent knocking at the door. It hadn't managed to wake Sam yet, so Jason got out of bed carefully and opened the door. Another bellboy stood at the door. He held a large basket of fruit and a bottle of champagne in a bucket of ice.

Jason hushed the man and stepped out the door. He squinted in the bright sunshine. The man handed Jason the basket. It was heavier than Jason would have guessed. Jason pulled a dollar bill out of his wallet and handed it to the man. He then carefully opened the door to keep it from creaking and placed the basket on the breakfast table. The cellophane crinkled lightly as Jason slowly opened it. He pulled the card out and read it.

"Congratulations Samantha and Jason! Best wishes for a fruitful life." All of the EnviroCon Project team members had signed the card. Jason smiled at the thought. He set two champagne flutes next to the basket and pulled the champagne bottle out of the ice bucket. Crystal, he was impressed. Jack must have arranged it. He pulled off the protective wire, put a towel over the cork and opened the bottle as quietly as he could. The towel muffled the sound of the cork. Sam barely stirred. He poured the champagne into the flutes, put them on the side of the bed next to Sam and crawled into bed next to her. He kissed her cheek and then the nape of her neck. She squirmed, stretched and opened her eyes.

"Hmmmmm, what time is it?"

"Time for champagne, Mrs. Graham." Jason picked up both glasses, kissed Sam and then, after she sat up, handed her a glass. He then raised his glass.

"To my beautiful wife, to whom I am eternally grateful for having married me." Jason took a sip of champagne and then kissed Sam again.

"Oh, Jason, I love you." Sam wrapped her arms around Jason and drew him toward her. They made love quickly and passionately. Afterwards, their energy renewed, they decided to go exploring. They had seen very little of Casa De Campo when they had driven in earlier that day. The weather was pleasantly warm, in the lower eighties, even though it was late in the afternoon. They walked from their bungalow through several walkways canopied with tropical vegetation, flowers and orchids. When they got to the lobby, they obtained information about the various scheduled activities and restaurants located at the resort. The first night they decided to eat the Luau at the Polynesian Palace. The desk clerk confirmed that there would be a band and dancing to accompany dinner.

Between organizing the start-up for the Project Runaway and planning their wedding, this would be the first time in several months that they would have any real time together. They walked over to the resort's beach and surrounding cove. They took off their shoes, got rumrunners from the beach's tiki bar and took a leisurely walk along the beach. The sand, already cooled from the heat of the day, felt good wiggling through their toes.

Surprisingly, the beach was not very crowded. A few couples sat reclining in chairs sipping drinks and several people were swimming in the cove, but to Jason and Sam, it seemed as if they had the beach to themselves. They walked arm in arm down the beach to the end of the cove and then turned around to go back. The sun was just setting over the ocean.

Jason stopped. Sam turned back and looked at Jason. "Why'd you stop?"

"I just wanted to take a long look at this scene so that I can remember it."

"Jason, I didn't marry you for your keen sense of romance. Aren't you laying it on a little thick?" Jason smiled and suddenly lunged forward, tackling Sam. "I needed to distract you so I could do this." They kissed and then Sam grabbed a handful of sand, threw it at Jason and jumped up.

Running down the beach, she shouted over her shoulder, "That's for spilling my drink!" Jason jumped up and ran after her, laughing. He finally caught her and stopped her, conveniently in front of another bar.

"Aren't you going to buy me another drink, Mr. Graham?"

"Why, of course I will." Jason got each of them another frozen rum drink and they sat down in a couple of deck chairs to catch their breath.

"You know, you misunderstood me back there on the beach when I stopped to admire the sunset. While it's beautiful and while I can't help but be romantic with you around, I was thinking about how much I'll miss all of this."

"What do you mean?"

"We're both going to be extremely busy over the next few years. I doubt that we'll have an opportunity for another real vacation. Even if we do, the Caribbean will not be the place to go. Hurricanes will ravage most of the resorts over the next few years. There won't be anything to come back to. That's one reason I wanted to come here on our honeymoon."

"Let's try not to think about the future now. Let's just enjoy the present." Sam leaned her head against Jason as they sat, sipped their drinks and watched the sun set.

⚓

MARCH 13, 2007.
3:20 P.M. EST.
THE CENTRAL ATLANTIC.

The pressure in the center of the storm had dropped further as the storm continued to gain strength and to organize. Sustained winds had reached 62 miles per hour and an eye was beginning to form. The storm would have already been given a name, but a new mail boy had just been hired at the National Hurricane Center in Miami. He didn't know that Emanuel Ortega was on vacation and that he should be delivering the satellite printouts to Samuel Weissman instead. Because hurricane season was over and because Mr. Weissman was not in the habit of receiving the printouts, he didn't miss them. Furthermore, the storm was located outside of the principal shipping lanes and air traffic patterns. No unusual reports had been received.

That would soon be changing. The storm intensified and moved towards the Caribbean. Given its current course and speed, it would be within view of the East Coast weather satellites within the next twenty-four hours and would hit the Lesser Antilles within forty-eight hours.

⟡

7:45 P.M. EST.
ALEXANDRIA, VIRGINIA.

Natalie was on her third cigarette, sitting at the bar in Alexandria when a man tapped her on the shoulder.

"Excuse me. Do you have a light?" Natalie turned around quickly on her barstool. The man was holding a cigarette between his fingers.

"What did you say?"

"I asked for a light."

"Sure." Natalie reached into her pocketbook to pull out the small plastic lighter. Her hand brushed briefly past a cold piece of steel before reaching the lighter. She lit it, but her hands were shaking so hard that the man couldn't light his cigarette. He grasped her hands with his, steadied them and lit his cigarette. While the bar was dark, from the light cast by the flame, Natalie could see that the man had brilliant blue eyes and sandy blond hair. He was also about six feet two inches tall and very attractive.

"My name's Bill. What's yours?"

Natalie answered without looking at him.

"Do you mind if I join you for a minute?"

Natalie turned to face him, deciding he was the type that would require a firm brush-off. "As a matter of fact, I do. I hate to be rude, but I'm waiting for somebody."

"I understand. Maybe I'll see you around here again. I come in fairly often."

"I doubt it."

"Thanks for the light."

"You're welcome." Natalie turned away from the man. She

watched him walk back over to a table where another man was sitting. She wanted to make sure she was rid of him before Kreschenko arrived. Why did he always make her wait? Didn't he know that she was taking a risk coming out like this? Finally she noticed him come in the door. He was alone. He walked over to a booth and sat down. She snubbed out her cigarette, paid for her drink, picked it up and walked over to meet him.

"Why, good evening, Natalie. You are looking lovely as usual." Natalie thought to herself, why did he always play these games with her. She felt as if he was laughing at her. She tried to ignore the remark.

"Why did you call me again? You know how tight security is."

"I know. I know. Do not worry your pretty little head. I'm used to dealing with it. I'll handle it."

Natalie hated his smug attitude. He didn't know Michael Barnwell as well as she did. Besides, ever since that man, Smith, had taken over security, she wasn't able to find out much from Michael. "I hope you're right." Natalie paused to sip her drink. As usual, Kreschenko was in no hurry to start the discussion. Natalie had canceled a date with Michael as a result of Kreschenko's last-minute call.

"What do you want now?"

"Why do you think that I want something? I enjoy your company."

"Cut the crap, Kreschenko."

"Ah, well, if you insist, it will be all business. I need to get copies of the program your firm developed to analyze the NASA database."

"I know. But it's harder than you think. The program is copy protected and I can't break the security code."

"But you've got to." Kreschenko thought about Greschko sitting in the office in Moscow.

"You've got to get the KGB files on my mother and father. You promised it months ago."

"There are," Kreschenko paused, "difficulties in Moscow."

"It can't be any more difficult than what I'm facing."

"I didn't say it was impossible, just difficult. But to show my good faith, I brought you the original indoctrination papers your parents signed." Kreschenko pulled a short manila folder out from under his jacket and handed it to her. She opened it and pulled the papers out. They were in Russian, which she did not read, but they looked official. The seal of the KGB was on the bottom of each page and there was a picture of her mother and her father at a young age attached to each file.

"Now, if you can manage to get me a copy of the program, I should be able to get you the case file on the information your father leaked to me."

Natalie smiled slightly. "I have managed to print out a few pages of the program. While I'll never be able to copy the program onto a disk, I have figured out a way to print the lines of code out a few pages at a time." Natalie looked down to dig into her pocketbook. As she did, she did not notice the sandy blond headed man with the blue eyes walk past her booth. He bumped into somebody as he passed by, turning his body towards her and Kreschenko. He flicked his wrist almost imperceptibly and then walked on back to his table. He leaned over and said a few words to the man he'd been sitting with and then walked out the door.

Natalie found what she'd been looking for and pulled a single sheet of paper out of her purse. "Here's the first page of the printout. I've got another forty pages hidden away at a locker in Union Station. We can go get them now if you like."

She handed the paper over to Kreschenko. It was covered with lines of programming code. Kreschenko had no idea what it meant, but it looked valid. "Yes, I'd like to get it now."

"Let's go." Natalie stood up to leave. Kreschenko grabbed her wrist. Hard.

"But you'd better be playing this one straight, Natasha. Or I'll deliver that file to Trey Wilson rather than you. No more crap like that first Report that you gave me." He released her wrist.

"I told you, that was an accident. I forgot to tell you not to copy the Report."

"There had better not be another accident like that again."

Natalie glared at Kreschenko while massaging her wrist. She picked up her purse, feeling the extra weight within it. "Don't worry. There won't be."

$\maltese$

Michael Barnwell was finishing up with his tai chi routine when his beeper went off. He dried off his face and picked up the phone in the locker room.

"Michael Barnwell."

"Mr. Barnwell, you've got an urgent call from a Mr. Smith."

"I'll be in my office in a minute to pick it up." Michael picked up his athletic bag and headed for his office. When he got there, he picked up the phone.

"Michael Barnwell speaking."

"Michael, Hamilton Smith here. We've got final confirmation that one of your players is dirty."

Michael frowned. He was surprised. He knew all of the team members and didn't anticipate a leak from his end. "Who is it?"

"Sit tight, I've got a courier on the way over with pictures and a transcript. But it's bad. I've initiated an immediate action." Michael knew what that meant and didn't like it.

"But you can't do that with one of my people without my reviewing the facts."

"Look, Michael. This line may not be 100% secure. Don't say anything more. You'll understand when the package arrives. Goodbye." The line went dead.

Michael hung up. His forehead was wet with sweat. He began to think about exactly what Hamilton had said and who might have been dirty. He called the receptionist and told her to bring a package down to him immediately when it arrived. Nakamura and Wayneright were the likely targets, but he'd had them monitored closely. He doubted that Hamilton Smith would have picked anything up on them. The reports he'd seen on them had been clean.

Who had been acting unusual in the office? Finally it dawned on him. How could he have been so blind? He picked up the phone and dialed a number. The phone rang four times before it was picked up. "Hello."

"Natalie, thank God you're..."

"You have reached..." Michael slammed the phone down.

✦

8:15 P.M. EST.
WASHINGTON, D.C.

Kreschenko sat in the front seat of Natalie's car. His thoughts wandered to the warm fire he would build at his new dacha and the large-chested secretary he would share it with. He looked over at Natasha. She had a nice set on her, just like her mother. The supple flesh of her cleavage invited him, but he held back. Perhaps he'd force the issue with her after he drained her of useful information. It would be fun. Her mother had been good in bed; he didn't doubt that little Natasha would be as well.

✦

8:16 P.M. EST.
WASHINGTON, D.C.

Natalie nervously smoked another cigarette, trying to cover the odor of sweat and cologne that always seemed to permeate Kreschenko. She'd decided to take a chance. She'd eliminate Kreschenko and hope that his boss wouldn't release the information on her father and that she wouldn't get caught.

She'd gotten a gun from Michael's apartment that he'd shown her one night. It was untraceable according to Michael. She'd taken it from his apartment several nights before and had left it in her car. To make sure that he wouldn't notice it was missing, she'd invited him to come over to her apartment for the past several nights. When Kreschenko had gotten the message to her earlier in the day, she'd made up her mind.

Several foreigners had been killed in the vicinity of Union Station over the past few months. She would make a wrong turn after crossing the Memorial Bridge and would come into Union Station through the bad end of town. Her car would stall and she'd get out to look at it. When Kreschenko got out to taunt her, as he inevitably would, she'd shoot him, take his wallet and the paper she'd given him and leave the body. The police would treat it as another random killing of a foreigner. She'd throw the gun and his wallet in the Potomac and then confess what she'd done to Michael. She hoped that he and the others would understand, but even if they didn't, they'd never go public. They'd have to help her cover it up.

When she turned onto the Memorial Bridge, leaving Virginia, traffic was light. Few people entered this part of D.C. after dark. Natalie didn't notice the couple in the shadows along the edge of the bridge or the reflection in the scope of the rifle just before it fired. The ice bullet tore a hole through the left front tire of her car, causing a blowout. Natalie lost control of the car. It careened off one guardrail and then broke through another to plunge into the icy waters of the Potomac.

Natalie banged her head on the steering wheel and lost consciousness. Kreschenko did not. He screamed as the car plunged into the water. He fumbled at his seat belt as the waters closed in over the roof of the car. Darkness quickly closed around him. He managed to release the seat belt just as the car settled on the bottom. In his panic, he tried to open the door, but could not. He finally realized that the water pressure on the door prevented it from opening. He could hear water seeping into the car and his feet began to get wet. The water must have been coming in fast because the water quickly reached his lap.

As he grasped the door handle, he heard a noise. He thought it must have been the car settling in the mud. The water was up to his neck. He took a deep breath. Just as he pulled on the handle, the door opened and water flooded in. He remembered hands pulling on him and a sharp pain in his neck before blacking out.

❖

10:45 P.M. EST.
NORTHERN VIRGINIA.

Kreschenko awoke with bright lights glaring in his face.

"So you're awake, Vladimir."

"Who's that?"

"Never mind. We've got a great deal to talk about."

Vladimir struggled, but couldn't move. "Am I in a hospital? I was in an accident."

"It was no accident, Vladimir, and you aren't in a hospital. Now, what information did she give you?"

"Who? What are you talking about?"

"You know who."

"You mean the girl in the car?" No answer.

Kreschenko bit down on his left molar.

"Sorry. We've removed your escape plan. And you will talk to us." Kreschenko felt a jab in his left arm. "Oh, don't worry. We're not going to do this the way the KGB taught you. We have drugs to take care of that for us. But you might wish we were doing this the old fashioned way. This new drug is very effective. We will get all of the information out of you that we want. Unfortunately, it will leave you a babbling idiot."

Kreschenko struggled against his bonds. "You can't get away with this! This is the United States, not Russia."

"You're right. But who will find out? We can get away with whatever we want as long as nobody finds out. And, let me assure you, nobody will."

Kreschenko struggled against his bonds but began to feel very dizzy. His eyes clouded over.

"Let me know when he's ready for questioning." The man in the grey suit left the interrogation room and entered the room next door. Hamilton Smith stood waiting for him.

"What did you find in the car?"

"We got this from Kreschenko and this from the girl." The man handed Smith the folder that Kreschenko had given Natalie as well as the computer printout that Natalie had given him. Smith glanced at them. "I think I know why she did it now. Was there anything else?"

"Yes, there was. We found this in her pocketbook." The man dropped the gun on the table.

"It looks like the little lady might have had a present in store for Kreschenko. Too bad. Did you take care of her?"

"She was unconscious when we got there. We injected the alcohol into her bloodstream. It will go down as a DUI accident, as planned. She was upset about her mother's death, got drunk, had a blowout. An unfortunate accident."

"Good. When you get all you can out of Kreschenko, get rid of the body."

"Yes, sir."

The man left. Smith wrapped up the evidence and took it with him. He'd have a little discussion with Michael Barnwell. He owed it to him.

CHAPTER 13

"What do you mean, you haven't seen the data yet! We have a category two hurricane bearing down on the Central Caribbean and no warnings have come out of this office! Goddamn it! I had to see the report on the morning news. Do you have any idea how this makes us look?"

Samuel Weissman cringed as Director Torrence excoriated him. "But it's not the season...."

"I don't give a damn. And I don't give a damn that this isn't your primary responsibility. Get your ass moving. We've got a crisis to deal with. I'll deal with you later." Director Torrence stormed out of the room.

❖

MARCH 15, 2007.
LA ROMANA, DOMINICAN REPUBLIC.

Jason and Sam had decided to go deep sea fishing that morning.

They'd booked the trip the night before and got up to eat an early breakfast before leaving. A strong wind greeted them as they walked through the gardens to the restaurant. Since neither of them suffered from seasickness they didn't plan to cancel the trip.

After eating breakfast, they strolled over to the lobby to catch the tram to the marina. The marina was located outside of the Casa De Campo compound in the heart of the adjoining town of La Romana. The drive took approximately fifteen minutes. When they got there, the thirty-three foot boat was moored to the dock. They didn't see the captain, so they climbed on board and took a look around. The rods were stored in the cabin and the boat was not ready for their charter.

They got off the boat and walked toward a small building built on the dock with a Casa De Campo sign hanging over its door. They walked past a short, burley haired man with leathery skin. They nodded to him and he nodded back. When they got to the building, the door was locked. They looked back over their shoulders and saw the burly man walk onto the boat they'd just left. They went back to talk with him. He certainly had the look of a charter captain about him. He had clearly lived on the water for most of his life.

When they got there, he had gone below deck. "Señor?" No answer. Jason cleared his throat and spoke louder, "Señor!" The man opened the door to the cabin and popped his head out.

"Sí."

"Are you the captain of this boat?"

"Qué?"

"I said, are you the captain of this boat?"

"Señor, yo no hablo Inglés."

Sam nudged Jason. "Jason, my Spanish goes way back to high school, but I know that he said he doesn't speak English."

Jason turned to Sam and smiled, "I figured that one out. Do you remember enough to ask him about our trip?"

Sam shook her head. "No, I barely remember how to count to ten in Spanish."

Jason looked at his confirmation slip that the resort had given him. It listed the name of the boat and the name of the captain. Jason put a foot on the gunwale of the boat, asked the captain in English whether he could come on board and watched his reaction. The captain signaled him to come aboard. Jason stepped down the three-step ladder into the cockpit of the boat.

Handing his confirmation slip to the captain, he said, "Do you know what this is?" The man scratched his head and then nodded. He spoke quickly in Spanish, but neither Jason nor Sam could figure out what he was trying to say. Finally, they managed through hand signals to ask the captain whether he would take the boat out that day to go fishing. The man shook his head vigorously, spoke several words in Spanish and then pointed back up the river rather than out to sea.

While frustrated, Jason got off the boat, took Sam's arm and left to go back to the hotel. No taxis were in sight and they had to wait fifteen minutes for the tram to arrive to take them back. Sam took things in stride, but Jason's mood blackened as they waited. The trip back to the hotel was uneventful, but Sam noticed and pointed out to Jason that a number of the shopkeepers along the route were closing their shops and boarding them up and that all of the artists that had lined the roads selling Haitian primitive art were gone.

Jason walked into the hotel lobby ready to have it out with the hotel manager. He asked the desk clerk for the manager, but she pointed to a sign that Jason had not noticed on his way in. It said, "Meeting regarding the hurricane in the Tiki Lounge at 10:30 a.m. All guests are strongly recommended to attend." It was 10:20.

Both Jason and Sam looked at each other. A hurricane in March? "Well, Jason, it looks as if the prediction you made on the beach a couple of days ago is going to hit home earlier than you expected. That's probably what the captain meant when he pointed up river and why the shopkeepers were boarding up." Sam didn't look happy. She waited for Jason to say something but noticed him staring off to one side in thought. "Jason, did you hear me?"

Jason nodded absentmindedly "uh-huh" and then turned to face Sam. "Yes, dear, I heard you, but I was thinking. You go on to the meeting and find out everything that you can. I've got several things to take care of."

"But..."

"I'll explain later. If the meeting breaks before I meet you, meet me in our room." Jason gave Sam a quick kiss on the cheek and watched her head for the Tiki Lounge. A gust of wind caught her hair, causing it to catch the light with a shimmer.

Jason tried to dial directly through to The States but couldn't get a line. He tried the hotel operator, but she said that the lines were busy. He told her to keep trying the number he'd given her and to try him in his room and page him if she got through. He then called the La Romana airport. The one flight they had leaving for San Juan and then Miami was booked. The flight for the following day was also booked, but the attendant told Jason that she doubted the flight would take off because the night's inbound flight had already been canceled due to the hurricane. When Jason asked, she said that she didn't know the status of flights out of Santa Domingo; her computer was down.

Jason then called the airline in Santa Domingo directly. Yes, they did have a flight out at 3:30, but it was sold out. The 5:15 to Miami was also sold out, but there was a chance that the flight would be canceled. In fact, all flights that day were booked solid. He could try standby, but there was already a long list. Jason thanked her, but declined standby and hung up. The last place he wanted to be if a hurricane struck was in the middle of an airport built along the coast.

Jason hung up the phone and went to the gift shop. He spotted several backpacks hanging on the wall. He pulled two of them down and put them on the counter. He also bought up as large an assortment of crackers, candy, soft drinks and water as he thought he'd be able to carry. The clerk looked at him as if he were crazy, but rang up the charge on his credit card.

Jason then went to the front desk and cashed one thousand dollars worth of traveler's checks. He carried his purchases back to the room and locked the door. He looked out the window. The skies remained clear, but a strong breeze was blowing. He was glad that their suite faced the mountains rather than the ocean, but, even so, they were only a few thousand yards from the ocean.

Jason turned on the television, something that he and Sam hadn't done the entire trip. He was surprised to find that the resort had access to CNN. He began to pack. He first put the food and drinks that he had bought into the backpacks with one change of clothes to cushion the load and with several rolls of toilet paper. He knew that if they got caught by a hurricane of any size in a third-world country, even in a four star resort, it might be days before they would get food or water and could be a week before they would be able to leave. He then packed the articles of clothes and other things that they would least like to leave behind in one suitcase and everything else in another. Just as he was zipping up the last suitcase, Sam walked into the room. She looked around the room and saw what he'd been doing.

"One reason I married you, Jason, is that you always know what to do." She gave him a hug and a kiss. "Things look pretty bad. While the resort manager downplayed the situation, we are going to get hit by a category two hurricane tonight. The eye of the storm is projected to pass within twenty miles of the resort."

"What did he suggest?"

"He recommended that everybody stay at the resort. He assured us that the hotel was built in the '90s to survive much stronger storms and that all of the guests could stay in the hotel tonight. But, for security purposes, he recommended that all guests store their bags in one of the conference rooms. He also said that several buses were available to take people to Santa Domingo. Most of the guests are going to try that."

Jason thought back over the several-hour trip up from Santa Domingo. There were only a couple of small towns between La Romana and Santa Domingo. He would not want to get stuck anywhere along the way. There would be very little shelter. "I, we don't want to do that."

"I agree. Anyway, he said that they would hold a hurricane party in the new hotel for the guests that wanted to stay. What do you think?"

"Well, I haven't been able to get through to The States, but I hate to stay here if we are hit. Unfortunately, I don't think that we have much of a choice. I've called the local airport and the Santa Domingo airport; all flights out before the hurricane are booked up. Neither of us speak the language, so I don't think that we should consider leaving the resort. Our only choice is to stay here. But I can't help but remember what happened when Emily hit Jamaica in '02." Sam shuddered. She had read about the rioting that had occurred after the hurricane had struck and how several dozen foreign guests had been beaten and robbed by the hotel staff; several died. The U.S. government finally sent the Marines in to restore order and to rescue U.S. citizens from the rioting.

"My thought is that we camp out right here. I don't want to be with everybody else if the storm hits. While I doubt that we will run into another Jamaica situation, I hate to take the chance. Let's make a show of moving our bags out of the room and storing them in the main lobby. That way, the staff will think we've moved out of the room. We can keep these backpacks and the one bag here."

"Do you really think that's necessary?"

"No. But why take the chance? We'll be as safe here as if we were in the ballroom when the storm hits, maybe safer. And if there are any "problems," I'd just as soon be here rather..." Jason stopped because CNN had switched over to their reporter at the National Hurricane Center in Coral Gables. Jason could see a satellite picture of the storm in the background. It had a well-defined eye and looked as if it were about to hit the island of Hispañola, very close to their location. The report confirmed that the storm was strong, was moving fast and that landfall was anticipated in three to five hours near the town of La Romana in the Dominican Republic. Yes, it was very late for a hurricane, but out of season storms were not unheard of.

Jason shut the television off and they moved their one suitcase to the lobby. They were surprised that they could still order a meal. Evidently the manager had a firm handle on the staff and had not let the staff evacuate. Jason and Sam both ordered roasted pork, a resort specialty, potatoes and assorted vegetables and ate well. Very few people had thought to take the precaution of eating before the arrival of the storm.

After eating, they decided to take a walk to the beach one last time. The wind had picked up and the sky was spotted with clouds, but it had not started raining. They walked around a mob scene in front of the lobby where people were scrambling to get on the buses heading for Santa Domingo. Jason just shook his head. He hoped that they would make it. On the way to the beach, Jason noticed one hardy soul teeing off on the sixteenth hole of the Teeth of the Dog, into the growling wind, gripping a lit cigar between his teeth. He could only smile and hope that the man finished his round.

The surf had already begun to pound the beaches and the clouds were building up on the horizon. Despite the situation they were facing, Jason couldn't help but think how beautiful Sam looked walking on the beach with the wind sweeping her hair back. They walked along the beach in silence until the rains began. While the wind alone had not bothered them, the light drizzle that had begun to fall stung with the force of the wind.

They ran back toward the hotel. It had started to get dark due to the clouds. Just before entering the hotel, Sam stopped. "Jason, do you hear anything?"

Jason stopped, but he couldn't hear anything over the wind. "No. Let's get out of the rain." Jason turned, but Sam grabbed his arm.

"No, wait!" Sure enough, Jason did hear something. It sounded like a jet. Jason and Sam both turned towards the sound at the same time. They saw a small jet swaying in the wind. Jason was just thinking how that guy must be crazy when he heard Sam say to herself, "My God, that's dad's plane! Jason, that's dad's Falcon. It's got to be. Who else would be landing in a hurricane?"

"Are you sure?"

"Yes, I saw the company logo on the tail before it dropped below the trees." They ran off in the direction of the resort's small airstrip. Jason and Sam had seen it the day before when they had played golf. It had been built in the middle of the golf course, and on one hole they actually had to drive over the runway. Not being much of a golfer, Jason had bounced his ball across the runway.

When they arrived at the runway, Jason and Sam could see somebody arguing with a uniformed guard next to the small terminal building. Their caddie had bought several beers for them at the building while they had been playing golf. When they got closer, Sam smiled. She leaned over and yelled in Jason's ear "It's Frank Osborne, my father's pilot." They walked over to the men and almost reached them before they were noticed. Both men turned in surprise. Frank smiled as Sam waved to him. He said something to the guard and then walked over to them.

"Why, Miss Samantha, how are you?"

Soaking wet, her clothes pasted to her body by the wind and rain, Sam just smiled and gave him a hug. She then turned to Jason, holding Frank's arm. "Jason, I'd like to introduce you to my father's pilot, Frank Osborne. Frank, this is my husband Jason."

They shook hands. "I hate to spoil the introduction, ma'am, Mr. Graham, but we'll be much more comfortable on board the aircraft." Jason didn't argue with that. Frank stopped to say something to the guard and then walked them onto the plane.

"Miss Samantha, Mr. Graham, I suggest that you fasten your seat belts. It seems that we are going to have to take off in a hurry. It may be a bumpy ride." Frank turned and began to crawl into the cockpit. Jason jumped up.

"What do you mean? We're not ready to go. Our bags..."

Leaning back over his seat, Frank answered while the copilot warmed up the engine. "Well, Mr. Graham, it's this way: It seems as if that guard and I didn't see eye to eye. He says he's going to impound the plane until after the storm. I convinced him that I needed to get some paperwork out of the plane and that I'd meet him in the building. If you go back for your bags, we won't be able to take off. I hope he doesn't hear the engines over this wind, because I have a feeling he might just use that gun he carries." Frank turned around and revved up the engines. Jason and Sam quickly strapped themselves into the back seat of the jet.

✛

MARCH 16, 2007.
MIDDLEBURG, VIRGINIA.

Jason awoke with a start. It took him a minute to realize that he was next to Sam in her room at her father's house in Middleburg, Virginia. He still felt a little shaky after their flight. While he'd never had a fear of flying, he had a great deal more respect for the air than he used to. Jason had barely managed to keep from throwing up after takeoff. At one point he felt that the little jet was about to be slammed back into the ground. Fortunately, Frank Osborne angled the plane away from the hurricane to get out of the rough weather fairly quickly. Jason and Sam took advantage of the ample stock of liquor that Mr. Whitlock maintained on the plane.

After the landing at Dulles International they both thanked Frank Osborne profusely and were grateful that he had radioed ahead to have Mr. Whitlock's chauffeur pick them up and drive them to the Whitlock estate. Mr. Whitlock had already retired when they arrived according to his butler, but Jason decided that Mr. Whitlock had done so for their convenience, knowing that they would be exhausted after their ordeal.

Jason flipped on the television to a news program and went to the bathroom. He was brushing his teeth when he heard the weather come on. He walked back into the bedroom to watch and noticed that Sam had propped herself up on the pillows. They both watched as the initial scenes came in from the Dominican Republic. The eye had passed over the coast just to the west of La Romana. Scenes from the news helicopter showed the devastation along the coastline. The chopper actually flew directly over the Casa De Campo Resort. While the hotel was still standing, numerous of the other buildings had been knocked down and trees were down everywhere. The reporter indicated that several dozen people had already lost their lives and that power and water were out for a large area of the country.

The storm had lost a great deal of intensity when it had hit the mountain range in the center of Hispañola and had been downgraded to a tropical storm. In addition, a cold front moving down from the Midwest was expected to prevent the storm from intensifying or from threatening any portion of the U.S. mainland. Officials at the National Hurricane Center confirmed that the storm was a fluke and did not anticipate any other off-season hurricanes.

Jason walked back into the bathroom to rinse out his mouth and then returned to the bedroom. "You know, I wish they were right. But I am afraid that we've lived through something that we'll be seeing a great deal of over the next few years." Sam nodded her head but had somewhat of a wicked smile on her face. She pulled the sheets back to reveal that she had changed into a sheer black negligee. She patted the bed next to her, "We're still on our honeymoon."

Just over an hour later, Jason and Sam both made their way down to breakfast. When they walked into the breakfast room they were both surprised. Sam's father had gone all out for them. Evidently he had not forgotten that they were still on their honeymoon either. A large arrangement of flowers sat in the middle of the breakfast room table. Two places had been set next to each other. Sam's had a bud vase sitting next to her seat. A pot of coffee sat on the table as well as a pitcher of orange juice and a bottle of champagne. Jason pulled Sam's chair out for her, gave her a kiss and sat down.

Jason poured each of them a mimosa, handed Sam her glass and then raised his: "To my lovely wife, a prize so fair that I'd fly through another hurricane to be with you." As they clicked their glasses, the butler brought in two plates of eggs benedict, fresh cantaloupe and hash browns.

Just as they were finishing breakfast, Mr. Whitlock entered the room. Jason stood to greet him. "Good morning, Mr. Whit—I mean Stephen."

"Please, don't get up. I hope you've both enjoyed your breakfast. I had Kim prepare your favorite, Sam."

Sam rose and walked over to give her father a kiss. "Thanks, Daddy. And thanks for rescuing us from the storm."

"You're welcome. When I saw the storm developing and couldn't get a message through to you, I knew that I had to send Frank down after you. You might not remember this, Sam, but a large hurricane trapped your mother and I in Mexico back when you were a young girl. It was one of the most miserable events in my life. Knowing that I had the ability to get you out, I couldn't let you suffer the same fate, especially on your honeymoon."

"We really do appreciate that, Stephen."

"Now that you are both relaxed, I've got some bad news for you." Mr. Whitlock paused to pour himself a cup of coffee. "I don't know how to put this delicately, so I'll just tell you. Jason, your secretary Natalie was killed in an automobile accident several days ago. Her funeral was yesterday." Jason and Sam both looked shocked. "It's true. I saved the short article in the paper about the accident for you. It would not have made the paper except that the accident occurred on the Memorial Bridge. She was drunk, had a blowout and went off the bridge into the Potomac."

"Natalie, drunk? That's not possible. I've never seen her drink heavily." Jason turned towards Sam.

"I don't think she was the type, but you never know. After all, her mother had just died."

"I know, but still. I'd better call Jack. I need to let him know that we're safe anyway. If you'll excuse me..."

"Certainly, Jason. Use the telephone in the library."

Jason came back several minutes later looking paler than before. "Sam, I think we need to talk. Will you excuse us?" Sam looked surprised but stood up and walked out of the room to the library with Jason. Jason closed the door behind them.

"What's wrong?" Sam hadn't seen such a look of concern on Jason's face before.

"I think that Project security killed Natalie."

"What? That's ridiculous! Why would they kill Natalie?"

"You know, I listened to them talk, but I didn't really believe that they would do it. Michael Barnwell told us that people had been killed in the past to keep double-black projects secret. But I didn't really believe it. The new security man with the CIA also made a veiled reference to the actions that could be taken, but it just didn't hit home."

"But what did Jack say?"

"It's not what he said. It's the way he said it. He knew that we couldn't talk over an open line. Especially after the accident. But he did say that we needed to talk when I got back to town. I could tell what he meant. Damn it! I can't believe it. I have always had doubts about keeping this secret. Yes, I know we've got to. My own models show me that. But I've always been concerned about too much power being placed in our hands with nobody to answer to. If they can do this, what else will they do? You know, eventually, the people in this group are going to have a lot more power than they do now. How can we control it?"

"I don't know. But what choice do we have? You said yourself that we can't afford to go public. And besides, you're in charge of the Project."

"You know better than that, Sam. I'm merely a figurehead, an implementor. They didn't consult with me on this decision and they probably won't in the future. Maybe I should just walk away from it all. Knowing what we do, we could take action to protect ourselves. I'm sure we could live out our lives in comfort."

"But what about everyone else? What about our children?"

Jason looked into Sam's face. A lone tear was rolling down her cheek. "Sam, you don't mean..."

Sam nodded. "I'm not sure, but I think so."

Jason gave Sam a hug. "That's great news." But a frown then

crossed Jason's forehead. "But you're right. We can't just think of ourselves or," patting Sam's stomach, "of just our children. But what can we do?"

"Either way, I think we'll be better off with you spearheading the Project. Besides, I know Geoff Barnes well enough. He'll do the right thing."

"I hope you're right. But we're not the only ones on the Project. I trust Jack and Bob Ryan, but as things progress, the size of the organization will grow. And its secretive nature creates a great deal of room for abuse." Both Jason and Sam stood in silence. "Sam, I've got an idea, but it will entail some risk; especially in light of what happened to Natalie."

"What is it?"

"We need to create a check on the power of the Project organization, something to fall back on in the event that we lose control over the Project." Jason paused. "Your father was in politics. He knows the game. He also controls a number of corporations that will be receiving contracts in connection with the Project. With our inside knowledge and with the support of your father, we could create a private group with the strength and the knowledge to counter that of the official group. But we'll have to be very careful, especially in the early stages of the Project."

"I think I see what you're driving at. We could point my father's companies in the right direction with research and development to be in the perfect position to profit from the impending disaster. Given the profits and the expertise we can provide, the companies will be in a position to counter actions of the official group. It might work. It will be difficult. We'll be walking a tightrope, but it might work."

"You know, if you quit the Congressman's staff and come to work for your father, we won't even have to hide where the information is coming from. You know all about the Project already. We can convince the other members of the executive committee of the importance of you working for your father's companies. Your actions, and the actions of your father's companies, will be private acts. Nobody will be able to point their finger at the government if a private company undertakes any unusual research. And if the companies happen to be positioned when the time comes to let government contracts, you will be deemed to be an expert planner."

"Come on. Let's go talk this over with your father."

When they re-entered the breakfast room, Mr. Whitlock was finishing his breakfast, a half a pink grapefruit. He looked up as they reentered the room. "Stephen, Sam and I have a great deal to discuss with you."

Stephen Whitlock wiped his mouth with his napkin, pushed back his chair and stood up. "No time like the present. But, given the looks on both of your faces, why don't we move back into the library? It's more comfortable." Mr. Whitlock walked briskly into the library. Sam and Jason followed him. He took a seat in a large leather recliner, angled to receive warmth from the fireplace. He then picked up an Algerian Briar pipe and a worn brown leather pouch of tobacco and began to fill his pipe. Jason and Sam took seats on the sofa across from Mr. Whitlock.

Jason stood up, put his hands in his pockets and walked over in front of the fireplace. He stood there for a minute soaking up its warmth and then turned back to face Mr. Whitlock. It took Jason close to an hour to recount the discovery that the Earth was in the early stages of a runaway greenhouse effect and the steps that EnviroCon had initially taken. He then described the meetings with Congressman Barnes and the Vice-President and the establishment of the Project Runaway executive committee. He described the public disgrace that he was about to endure as well as his fears that the Project's security team had killed Natalie. When he finished, he sat down next to Sam.

Mr. Whitlock put his pipe down in a large malachite ashtray. "It sounds as if we are all in quite a pickle. And I agree with your concerns. Unbridled power has always led to abuse. I know Geoff Barnes and Tom Hammond, and I trust both of them. Despite their being politicians, their motives are sound. However, as you have correctly surmised, they will not have full control over the creature they created. The death of your secretary is a testament to that." He picked up his pipe and began to pack new tobacco into its bowl.

"They've got you walking a fine line, my boy. And they plan to keep you there. While both Congressman Barnes and the Vice-President have valid motives, you are the one that is out on a limb. Don't get me wrong. In their position, I would have done the same thing. You would, too. But what you need to do is to regain some of the control over the instrument that they have placed in your hands."

"Your idea of bringing me into the picture is a sound one. I'm old enough not to care what happens to me. But I am concerned about the two of you. I think that I can work with the R&D people to re-arrange some budgets to position us better for the future trends we will be facing, but there are limits. I do have shareholders to answer to. And while my control of the company is firm, it is not absolute." He smiled at Sam. "With Sam coming on board and with the information that you will provide, I'm sure that we'll work things out. You've given me the type of head start that any industrialist would want. We can do it!"

"Now, enough talk of gloom and doom. You are still on your honeymoon. Things are not so time critical that you can't finish it." Mr. Whitlock pulled an envelope out of his jacket pocket. He stood up and handed it to Jason. He then sat back down. "You'll find two tickets to Salt Lake City in there as well as a reservation for a suite at the Stein Erickson Lodge in Deer Valley. I'll have the chauffeur drive you to the airport this evening. I've heard that the snow's great. Relax, enjoy yourselves. We can work out the finer points of our plan when you get back."

"Stephen, come on. You've done enough already." Jason stood and attempted to give the envelope back to Mr. Whitlock.

"No, no, I insist. I can afford it and I won't take it back. Go on. Have fun. From what you've said, this may be the last time that either of you will have to relax for quite a while. Now go."

CHAPTER 14

The second of the Iranian terrorists convicted of killing the Saudi Crown Prince in last week's bombing was publicly hung today. Tensions have been building between the two nations since the fundamentalist government of Iran refused to condemn the bombing, instead, issuing a statement that the act was justified as an action against the "infidels" that had turned their backs on the teachings of the Prophet. Iranian mullahs have openly encouraged the overthrowal of the Saudi Royal Family to wrest control of Mecca from the "whores" to the industrialized West. The Saudi Air Force has been placed on highest alert to counter threatened incursions by Iranian aircraft.

Despite the arms embargo imposed on Iran last September, intelligence sources report that the Iranian military continues to gain strength through its dealings with the Central Asian Moslem Confederacy. Iran has stockpiled several hundred old Soviet-made intermediate range ballistic missiles and there is a great deal of speculation that the Iranian government has also obtained an undisclosed number of nuclear warheads from the Confederacy.

Oil prices on the spot markets around the globe have increased by ten percent since the assassination of the Saudi Prince, causing tremors throughout world financial markets.

◆

THE FARM.
BLUE RIDGE MOUNTAINS, VIRGINIA.

When Jason arrived at his office that morning, he pushed the stack of reports that always sat on one corner of his desk off to one side. He needed the time to prepare for the meeting of the executive committee scheduled for later that day. He was pleased with what they had accomplished over the past year. They had managed to stay fairly close to the goals that Jason had established for the Project, but the initial phases were winding up.

After firing Jason, EnviroCon completed its contract with NASA. Jeffrey Nakamura had done an exceptional job altering the NASA data. The results and the falsified data had been available to researchers around the world for several months. So far, the modifications made to the data had not been detected. While several scientists had expressed surprise at the results obtained from the data, nobody questioned its authenticity. Several environmental scientists had written reports indicating that the data refuted the fears that many people had held about global warming.

General Maxwell had moved forward with the reopening of the lunar base and had pronounced it ready for active operations. Congress approved additional funding for the Space Station. Congressman Barnes had delivered a number of key Congressmen to get the budget passed. But progress was beginning to stall. The next step was a big one and things looked bleak. They had to begin construction of the first of the space cities within the year. While General Maxwell had done wonders with the funds he had, he could not do much more and remain invisible. The elements needed to begin construction were in place. Now all they needed was a reason to build. The

Chinese had played into their hands and had continued to expand their presence in space. But with the election coming up, Congressman Barnes and the Vice-President both indicated that additional funding would not be possible.

They would be committing political suicide if they publicly advocated additional spending increases. As it was, Congressman Barnes' opponents had already made hay with his support of additional funding for space. Despite the activities of the Chinese, Congressman George Gallager argued for cutting the programs to feed the hungry at home. Why compete with the Chinese in space? Let them waste their money. What had the money spent on the expansion of the Space Station brought home to the public, he asked. Pretty pictures, to be sure, but wouldn't the money be better spent on job programs for the unemployed and on building houses for the homeless?

Unfortunately, the public was listening. After a fast start in the primaries, Congressman Barnes had already lost the status of front runner for his party. It looked as if he would not be the candidate to challenge the President, whose popularity had firmed up in the polls. Political pundits predicted that the President would be reelected. Despite growing evidence that the country was about to enter another recession, the public supported the President and his plan to increase funding of entitlements. His universal health care program remained extremely popular.

The President had rebuked the Vice-President for public statements he had made in favor of the space program. He would be taken off of the ticket if he didn't tow the President's line. The entire executive committee feared that the progress they had made would be lost in light of the growing recession and public support for the President. The Vice-President dared not publicly disagree with the President for fear of being removed from the party ticket. And the entire committee agreed that the Vice-President had to retain his position.

Progress could be made in secret, but not fast enough. Very few launches could be made from Vandenberg without public knowledge, even with the new Single Stage to Orbit Shuttle.

They needed something to move the public. Jason didn't have an answer. He hoped that one of the other members of the executive committee did.

Jason left his office and headed for the conference room. When he got there, Frances Crawford and Bob Ryan were already there. Bob Ryan came up to Jason to tell him that General Maxwell had run into a problem and couldn't make the meeting. Bob would stand in for him. Jack Watson arrived with Michael Barnwell. They both walked over to Jason.

"Jason, how are you? It's been a long time."

"I'm doing good, Jack. How is Michelle?"

"She's fine, although she still gives me hell about the way I treated you last year. You know, we had quite a row about it."

"I know, Jack. You told me last time we met."

"Well, she hasn't let up. I wish I could tell her the truth."

"I wish I could tell all my friends the truth, Jack. But it's just one of the sacrifices that we've all had to make."

"How's Sam doing?"

"Quite well. You know she's pregnant again."

"No, I didn't. Congratulations. I was very sorry to hear about her miscarriage last year. I wish we could have sent something, but we had to keep up appearances."

"Actually, Michelle did send us a very nice note. She didn't tell you?"

"No. But that's just like her. Well, good luck with the new one."

"Thanks, Jack."

❖

Michael Barnwell stared at the scene of the apple orchard displayed on the room's false window. He had missed the end of Jason and Jack's conversation. Instead, he had been thinking about Natalie drowning in the Potomac and shuddered. He should have guessed that something was wrong. Natalie had made the ultimate sacrifice due to his neglect. He hoped that in the little time he had left, he would be able to make up for his neglect. His stomach cramped and a wave of nausea passed over him. It took all of his discipline to keep from wincing. He excused himself and walked quickly down the corridor to the bathroom. He opened the stall and vomited into the toilet. He carefully flushed the toilet twice to make sure that no evidence of the blood in his vomit remained. He then washed his face in the sink, pulled a vial of pills out of his pocket, took one out and swallowed it. The pills were supposed to help with the nausea and the pain; but, recently, hadn't done much good. Michael composed himself and returned to the conference room.

Upon returning, he noted the Vice-President's Secret Service entourage mulling around in the corridor outside of the conference room. However, when they noticed him walking up, they quickly assumed their usual somber demeanor. He opened the door and quickly scanned the room. The entire executive committee, other than General Maxwell, had arrived. A Marine private had just set a tray of drinks and a pot of coffee down on the table located along the east wall of the room. Michael walked over to the table, got a ginger ale—his stomach could no longer handle coffee—and sat down next to Hamilton Smith. They gave each other a brief nod. The conversation broke up and people began to take their seats. Jason sat at one end of the table with the Vice-President seated at the other end. When every one had taken a seat, Jason began the meeting.

"Thank you all for coming to this meeting. We have made a great deal of progress over the past year and have managed to stay on schedule. General Maxwell has reactivated all of the automated facilities left at the lunar base and assures me that everything is progressing smoothly. The reactor is back on line and is processing lunar soil to extract oxygen and store it so that when we do return to the base, a supply of oxygen will be available. In addition, while one of the automated moles failed, the other one is functioning properly and has begun to excavate sites for the installation of new buildings. Once the mole has finished with the building sites, it will begin the process of clearing the several miles of track that the mass driver will require.

"We have also signed a contract with Whitlock Industries to develop an automated machine to process the lunar soil and mold it into a plastic that will serve as the basic construction material for the expansion of the lunar base. Whitlock Industries has already developed a new plastic that can be poured in place like concrete, has the tensile strength of steel and has extremely high insulation factors. Tests show that the lunar regolith contains all of the materials necessary for the plastic to be manufactured. The project chief believes that they will be able to devise a method to remotely manufacture, process and pour the plastic within a year. Once built, we can test the chambers to confirm that they are airtight and then cover them up with the excess lunar soil generated by the mole and by the plastic processing plant to provide radiation shielding.

"We can and will do all of this under the double-black funding elements of the program controlled by General Maxwell. With regard to the public elements of the program, the increased funding given to NASA and the expansion of the Space Station as well as the testing of construction techniques have progressed very well. Unfortunately, as several of you made clear in your reports, given the current political climate, funding will be cut in the next fiscal year. Despite the success of the Space Station and the potential benefits that we have and will continue to receive from it and despite the growing Chinese presence in space, President Wright has promised to cut funding to the program. Spending for the space station, as well as for several other science programs, is 'unnecessary' and will be cut to avoid tax increases.

"While I will not debate the economic legitimacy of the President's proposed actions, he has the public's support. Our projections show that he will be reelected by a slim margin. Even if he doesn't win, the Republican candidate—which, unfortunately, is not going to be Geoff Barnes—will take similar steps, although not as drastic as those proposed by the President. And, as we all know, regardless of the economic outcome of the President's actions, the steps that he proposes to take will cost our Project a delay we can ill afford. We need significant public funding in space. That's what we are here to discuss. The floor is open."

Bob Ryan cleared his throat and then spoke up. "General Maxwell requested that I relay his support for the proposition that we build up the Chinese menace again."

Sick and tired of the General's constant second-guessing of committee decisions, Congressman Barnes retorted: "Come on, Major Ryan! Even the General has got to understand that we've been outflanked! We tried it. The public is not ready for another space race. The Chinese and the Japanese have both done their homework. The Japanese quickly took the offensive and persuaded the public that they had no involvement with the Chinese space program. Given the number of newspapers, lobbyists and corporations they have influence over, it's no surprise that they defused our "leak". As to the Chinese, they have also managed to convince the public that they are not a threat or an evil empire bent on conquest. Besides, they are one of our largest trading partners. We couldn't succeed in portraying them as evil for long."

"Fears of inflation and a ballooning federal deficit, fanned by the likes of Congressman Gallager, have overcome the initial public support for a new and reinvigorated space program. What the public doesn't realize is that even with the increases we got last year, NASA's budget is still only about half of what it was during the height of the Apollo program. All the public looks at now is the total dollar amount in the budget, not the long-term benefits of the program. I'm afraid the General is sorely mistaken if he thinks we can raise the fear of a yellow tide in space. Please relay my thoughts to the General, Major."

"I will, Congressman. But what about the Russians? Can't we bring them into this?"

"No. They are a capitalist country now, and not a very strong one at that. The Japanese are a far more credible threat. No, we can't play the Russia card either."

"But didn't the Russians just launch another probe to Mars?"

"That's right. Evidently, the Russians are moving aggressively back into space. It looks as if Greschko has gotten a firm hold over the President and key members of Parliament. We all know the reason. Has anything leaked over there yet?"

Hamilton Smith sat forward in his seat. "No, sir, Congressman. We have conducted an intensive analysis of the actions taken by the Russians. It is highly unlikely that a leak will develop from their end. Only Greschko and a few of his top aids know the true reasons behind their push toward Mars."

"What about contacting Greschko directly, have we considered that?"

"Yes, Congressman, but we deem it to be inadvisable at this stage. If an opportunity arises, we will let the Committee know."

Jason scanned the room. The Vice-President seemed content to sit back and listen to the comments of others. Unfortunately, nobody was jumping forward with any ideas. "Does anybody have anything to add? Any other possible suggestions?"

Frances Crawford had been scribbling something on a pad in front of her and put it down. "I've done a great deal of thinking about this, gentlemen. We were bound to come up against this problem eventually. And given the past propensity of the public and our political system to avoid long-range investments, I'm somewhat surprised that we secured the funding we did last year. I consider it a testament to the political acumen of both the Vice-President and Congressman Barnes. However, there's nothing that they can do now."

"We need a justification for a major move in space. The public will not buy economic or political competition. We need tangible, near-term benefits. What does space have to offer us in the near term?" Frances paused. "Energy. Space has abundant sources of free energy just waiting for us to harness it. The problem has always been the cost of harnessing that energy." Frances used her light pen to activate the computer view screen. "As I'm sure all of you are aware, the originator of the idea to build cities in space, Dr. Gerard O'Neill, ran into the same problem that we have right now. How do you pay for the construction of the space cities and justify them on an economic basis? Construct massive solar power satellites to collect the free energy in space and transmit it back down to the surface of the planet."

"Unfortunately, energy costs have always been too low to make the cost of solar power satellites economically viable. However, the economics of energy are changing. The cost of energy generated by solar power satellites currently exceeds that of coal-fired plants by about ten percent."

"Basically, gentlemen, the cost of construction of Solar Power Satellites and ground-based receiving stations is not out of line with current energy costs."

"It won't sell." Frances looked in surprise at the Vice-President. "Even if the cost to construct the satellites were cheaper than ground-based power plants, the idea will not sell. It's too politically risky. Construction in space is too new. Even if feasible technologically, the project will be too easy for luddites like Congressman Gallager to attack. Unfortunately, America no longer has the vision it did when it built the Panama Canal or when it sent Man to the Moon for the first time. The public now expects nothing from the government but handouts."

"The idea needs a strong-willed President capable of drawing public support behind the project, as well as a legitimate and compelling reason to build the satellites. We have neither."

Jack Watson interrupted, "But the price of oil just went up $5.00 a barrel. If those prices hold, the cost of a solar power satellite system will be competitive."

"That doesn't matter, Jack. Oil went up because of the current tensions in the Middle East. Hamilton, correct me if I am wrong, but my last briefing indicated that the Iranians were about to back away from the current crisis and issue a private apology to the Saudi Royal family. But even if oil did maintain its current price, or go higher, the President will not move. There is no upside for him."

Frances Crawford was not satisfied. "But what if the crisis intensified? Couldn't you put a bug in the President's ear?"

"Yes." The Vice-President paused. "But it would take a very dramatic escalation in the Middle East to change his mind. And that's not going to happen. Is it Hamilton?"

"No sir. I believe that your reports are correct. The Iranians are being pressured by Syria, Iraq and the Central Asia Confederacy to back down. They will back down. Oil prices should return to pre-crisis levels within the next six to eight weeks."

Jason noticed that most of the people in the room shook their heads in disappointment. It was unbelievable to Jason, but they were actually hoping for a major conflict in the Middle East. And while it disgusted Jason, he realized that he too had hoped that the Middle East crisis might just be the catalyst they needed. After all, if Project Runaway failed, the people in the Middle East that would have died if the current crisis escalated would die anyway; they would just be given a few extra years, many of those in misery.

◆

Michael didn't pay attention to the balance of the meeting. He and Hamilton Smith had come to an understanding about what had to be done. And he was the one to do it. They met briefly after the meeting to finalize their plans. He and Hamilton Smith both knew that this was a decision to be made by them rather than by the politicians.

Michael returned directly to his apartment after the meeting to collect the materials that he would need. He pulled out a generic olive green duffel bag manufactured in Taiwan. He filled it with several articles of clothing, including desert camouflage fatigues. He placed a small nine-millimeter pistol made out of composite plastics that would get through any airport security system under his jacket. He would pick up the balance of the equipment en route. He changed into his old Air Force uniform, made one phone call, took a last look around his apartment and then left for Andrews Air Force Base.

As planned, Hamilton Smith had provided Air Force transportation for Michael to a small NATO air base outside of Rome. Once in Rome, Michael changed out of his uniform and destroyed it, made several telephone calls and then left to catch a freighter leaving the port of Ostia. The trip was rough. Michael threw up several times, the crew laughing at his sea sickness. He disembarked in Athens. There, he took a cab to a section of Athens frequented by foreign military personnel and checked into the Pension Tsalikopolis for the night as Major Jacques Lebow. Rather than heading for the bars, strip joints and whorehouses that flourished in this section of Athens, he went up to his small room to sleep. The next morning, feeling refreshed, he donned a French Major's uniform and took a cab to the NATO base. Once there, he gained access to the base using false NATO identification papers. He signed up for a military transport leaving for Cyprus.

Once in Cyprus, he left the military base and took a cab into town. He went to the Nicholaevich hotel, went up to the front desk, asked for his key and any messages. The concierge gave him the key and a brown envelope. Michael went up to the room, turned on the hot water in the bath, and opened the envelope. While steam filled the bathroom as the water filled the tub, Michael poured the content of the envelope out on the bed. The envelope contained a ticket in the name of Thomas Bainbridge, a British Passport with his picture in it as well as several pages of stamped entry and exit visas and a piece of paper with one name, El Hasat. Michael undressed, pulled a beer out of the small refrigerator in the room and climbed into the tub. It felt great. He sat there sipping his beer for about half an hour. When he got out, he did an abbreviated tai chi ch'uan routine, changed into jeans and a plaid work shirt, picked up his duffel bag, went down to the lobby and caught a cab to the airport.

The flight was uneventful. When Michael arrived at the Haifa airport, he passed easily through airport security. He took a cab to a small bar named the El Hasat and sat down on a stool at the bar. He ordered a bottle of Al Jacquime beer. By the time he had finished half the beer, he heard a shuffle behind him. He did not turn.

"It is very good to see you again, old friend, but on very

short notice." A middle-aged man with greying hair at his temples, and wearing jeans and a T-shirt sat down next to Michael. He carried a long, jagged scar along his left forearm.

"Shalom, Saul. It has been a long time. Too long. We have much to discuss. But, before we do, can I offer you a beer?" Michael signaled the bartender for another beer. He brought one over and set it down in front of the man that had joined Michael. They both took long pulls off of their respective beers, Michael finishing his. "Quite good. I'd forgotten how much I like your local beer. Can I get you another?"

"No, I have a car out back. Let us have one at my apartment." The man stood up to leave. Michael followed. When they got outside, Michael and the man crawled into a small Fiat and took off.

"I am sorry, my friend. The air conditioner is not working. We can no longer purchase freon."

"That's quite all right. The heat feels good to me." Michael rolled up his sleeves and stuck his hand out the window. They drove for fifteen minutes through a number of narrow, winding streets and alleys until they stopped alongside a stone building in an ancient section of town. Michael followed the man into the building. Children were playing in the front room. They walked down a short hall until they came to a door. They entered the back room. It had a simple bin and a wash stand in it. A curtain hung along the wall on the left-hand side of the room. Saul pushed the curtain aside and stepped into a small closet. He pressed against the concrete in one place and a panel slid to one side. A light came on in a stairway.

They walked down the stairs until they were below street level before the stairway ended at another door. Saul entered a code in a key pad next to the door and the door opened onto a well-lit corridor. Michael followed Saul down the hall and into a small windowless room with a round wooden table and wooden chairs in it.

Michael was a bit surprised at the room, but said nothing. He took a seat and waited for Saul to sit down. He did.

"Now, Michael, what is it that I can do for you? You didn't come all this way on such short notice for a social call. By the way, I thought you'd retired from the game."

"I had. But given the circumstances, I have no choice but to get back in the field. Besides, I've rather missed field work."

Saul laughed, "I'm sure you have, especially after that last mission we worked on together. You miss the blond nurse that looked after you while you recovered from the Iraqi bullet that you took for me."

"I didn't take it for you. I just tripped and got in the way." Saul smiled and nodded. Michael ignored his knowing look. "Okay Saul. Let's get started. I assume the room is clean."

"You must ask?" Saul leaned forward in his chair, his dark brown eyes staring into Michael's. Michael met his gaze. "What are you into Michael?"

"It's a long story. After I outline the operation, you'll understand. This is one operation where we cannot afford a leak. That's one reason that I came to you. Except for one highly placed individual, U. S. intelligence services know nothing about this operation. I'm flying solo on this one, old friend."

"Not anymore. You can count me in."

"Let me finish before you commit yourself."

"I'm committed. Now, go ahead."

✦

MAY 27, 2008.
HAIFA, ISRAEL.

Several hours before sunrise a small aircraft with no markings took off from an Israeli military base. It darted high over the skies of Jordan and Iraq at better than three times the speed of sound. Neither Jordanian nor the more sophisticated Iraqi radar detected it. Its active and passive stealth features rendered it invisible to radar. As an added precaution, the body, painted with a special light absorbing black paint was designed to be all but invisible at night.

The pilot spoke over the aircraft's intercom in English, "We are approaching Iranian air space."

One of the two men sitting in the rear of the aircraft removed his seat belt and stood up. He had to crouch slightly to walk the few paces to the aircraft's cockpit. He tapped the pilot on the shoulder and bent over. "You know the plan. Carry it out." He returned to the rear of the cabin and tapped the other man on the shoulder. They looked out the windows of the aircraft. The lights of Kuwait City shimmered off in the distance to the south.

As soon as the aircraft crossed Iranian air space, it circled around and began to descend. No image appeared on the Iranian radar. Just before passing over the Iranian border, at a point equidistant from two Saudi coastal radar installations, the pilot switched on the plane's electronic countermeasures to create the image of a small aircraft. The phantom plane dropped below the level that intelligence reports had shown the Saudi radar would lose a true aircraft before the pilot switched off the electronic countermeasures, causing the phantom to disappear.

Several seconds later the pilot used the intercom again, "We've just intercepted radio reports from both Saudi radar installations as well as one Iranian installation. Our ghost was picked up and then disappeared. They are trying to reestablish contact."

One of the two men in the back of the aircraft acknowledged the message with a thumbs up and then turned toward the other man. "It's not too late, Michael. We can still head back without being detected."

"No, Saul. I'm committed. How long until we reach the oil fields?"

"Just a few minutes. Let's get the equipment together." Both men were dressed in black with their faces painted black. Michael put a parachute on his back and Saul handed him a large and a small backpack. Michael strapped them onto the harness he wore over his shirt. He also picked up an old

Kalashnikov machine pistol, the favorite of Iranian terrorists. Saul also strapped on a parachute and picked up a Kalashnikov.

"What are you doing, Saul?"

"Well, you don't expect me to stand in the door of the aircraft without a parachute; do you? If we hit some rough air, I could get bumped out. Besides, even with our infrared sensors, given the heat, we might miss some Saudi guards. I might need to use this as you parachute to the ground." Michael nodded his assent. "Good luck, my friend." Saul embraced Michael. "Is there any message you would leave with me?" Michael shook his head. The pilot came back over the radio. "Two minutes to target. The drop area looks clear." Both men stood in silence by the door of the aircraft. They felt the aircraft slow to subsonic speeds. The drop would be made at just under one hundred miles an hour and the shock would jolt Michael, so he prepared himself mentally.

The pilot signaled. Saul opened the door to the aircraft. The wind whistled into the cabin of the aircraft, but the pilot managed to keep the aircraft from jolting. Saul embraced Michael again and then Michael jumped out. The wind battered Michael and then he felt a strong jolt as the parachute opened. He looked up but could barely see the chute. They had intentionally picked a moonless night and a black chute.

Michael didn't hang in the air for very long. He prepared himself for impact. He could barely make out the ground in the dim lights from the distant oil refinery. His knees absorbed the initial shock and he rolled. He quickly pulled in his chute and folded it. He then dug a hole in the sand and buried it. There was always a chance that it would be found. Michael heard a soft thud behind him and spun around, the Kalashnikov pointed in front of him. He saw nothing.

He listened. "Michael." He looked ahead into the darkness, a sand dune blocking the light from the refinery. He saw a shape and let it approach. He whispered. "Saul, what the hell are you doing!"

"I'm sorry, old friend, but I couldn't let you come alone." Michael glared at him but said nothing. After they finished burying Saul's chute, they headed off over the sand in the direction of the pipeline that led from the oil refinery to the coast. It should be approximately one mile from their location.

The walk over the sand took roughly fifteen minutes. They saw no signs of life, but, when the wind shifted, could faintly hear the sound of machinery coming from the refinery. When they got to the pipeline, they sat down. Michael opened the flap of the larger of the two backpacks to reveal a simple-looking panel with a keyhole in it. He removed a key from his pocket and inserted it into the device. The Soviets had designed the device for mobility and simplicity in use.

Michael turned the switch, hit a few keys and turned towards Saul. "It is done, my friend. Now all we can do is wait." Saul pulled a pack of Turkish cigarettes out of a pocket and lit one up. He shook the pack and offered one to Michael. He too pulled one out. Saul lit his with a match and then handed the match to Michael. He lit his as well. They both cupped their cigarettes out of habit, to keep the glow of the burning tobacco from showing. But they were not trying to keep hidden. Michael looked at his watch and signaled Saul. They both stood up and walked a short distance away from the pipeline. The Saudi guards should be passing this point in a few minutes.

They stood quietly and waited. The last phase of the plan depended on the timing of the Saudi guards. While it wasn't crucial, it would help if the guards arrived on time. They did. Saul saw them first and pointed down the pipeline towards the coast. The lights of the hummvee shined over a dune. Michael and Saul crouched down until the last-minute and then jumped up and began shooting. They shot out several tires of the hummer, stopping the vehicle, but made certain not to shoot the driver. The vehicle stopped suddenly as it veered into a small dune. Several shocked Saudis jumped out of the rear of the vehicle. Saul jumped up and ran towards them shooting wildly, shouting in Farsi "death to the Faisal infidels." The Saudis shot back, one managing to shoot Saul's leg out from under him.

Michael jumped up and ran towards the vehicle. Unlike Saul,

he didn't try to miss. He shot all three of them and silently approached the driver. Saul fired off several shots, letting Michael know that he wasn't dead and making the driver think that the battle continued. As Michael got within range of the vehicle, he heard the driver's desperate pleas for help. Michael waited patiently in the shadows. When Michael heard the acknowledgement and heard the driver mention Iranian terrorists, Michael took careful aim and shot the driver in the head. Michael smiled as the voice of the radio operator tried to raise the guard. The plan had worked perfectly, so far. He hoped that the balance of the plan would succeed.

He walked over to Saul's position and looked down at him. Several high caliber bullets had shredded the lower section of Saul's leg. It looked as if the only thing holding his lower leg to his knee was the skin at the back of his knee. Blood poured from the wound. "How are you, old friend?"

Saul coughed. "I've been better. At least, I don't feel anything."

"Well, it won't be long now. Would you like a cigarette?"

"Yes. They are in my pocket, but I can't quite reach them."

Michael reached into Saul's pocket and pulled the pack of cigarettes out. He pulled two cigarettes out of the pack, lit them both and put one into Saul's mouth. He drew deeply on his own cigarette. "Well, Saul. We did it." He did not hear Saul's response. The timer had run out on the device they had planted next to the pipeline. It exploded. Michael felt a very brief flash of heat before the blast vaporized him, Saul, several thousand cubic yards of sand around them and a large segment of the pipeline.

The fireball extended outward and upwards from the blast, rapidly expanding. It didn't reach the refinery, but the heat did, igniting the oil in the refinery moments before the shock wave from the blast blew the fire out and knocked down ninety percent of the structure. The fires reignited the refinery and that portion of the oil field directly below ground zero. The mushroom cloud rose rapidly over the burning oil field.

8:04 P.M. EST.
WASHINGTON, D.C.
BLAIR HOUSE.

The Vice-President finished his shrimp bisque and glanced around to see if his guests were ready for their next course, beef wellington, the Vice-President's favorite. Congressman Gallager had finished and was chatting amicably with Senator Thomas's wife. All of the others had finished except for Senator Gallager's wife. He could hear her slurp over the din of conversation. As usual, she was holding the dinner up. The President, campaigning in Texas, had asked the Vice-President to host the dinner to divine the views of key Congressmen and Senators.

Mrs. Gallager finally signaled the waiter to take her bowl away when the Vice-President noticed George Harper heading in his direction. George Harper leaned over and whispered into the Vice-President's ear. "Mr. Vice-President, something has come up. You are needed in the Situation Room."

He signaled George to lean over. "What is it, George?"

"Mr. Vice-President, there has been a report of a nuclear incident in Saudi Arabia."

"Son of a bitch!" All of the heads at the table turned towards the Vice-President. He stood up. "I'm sorry, ladies and gentlemen. You will have to excuse me." He turned to leave, knowing he would be questioned.

"Wait a minute, Tom. What's going on?" He should have guessed that Congressman Gallager would be the first to speak up.

"Unfortunately, George, I don't have all of the details yet. I will contact you as soon as I can."

"That's it?" Congressman Gallager wasn't satisfied.

"I'm afraid so, George. At least until I get a full briefing. I suggest you flip on the news. No doubt this will be on the air within the next ten minutes."

All of the guests began to stand up. The Vice-President's wife walked over to him. "Nothing to worry about dear, but I doubt I'll be back tonight." He kissed her forehead and followed George Harper and several other Secret Service agents out of the room and to the elevator. They went down to the tunnel that led to the White House.

When the Vice-President reached the Situation Room, several military officers were hovering around Harland Danielson, the National Security Advisor. "What have we got, Harland?"

"I just arrived myself, Tom. Major Anthony, can you give us a briefing."

"Yes, sir. Mr. Vice-President, at exactly 7:48 p.m. Eastern Standard Time, our early warning satellites picked up the flash of a nuclear detonation over one of the largest oil fields in Saudi Arabia. Preliminary readings indicate a relatively low yield, approximately twenty kilotons, but we will need to await further spectral analysis to be certain. Secondary fires in the oil fields and from the large refinery located 3.2 miles from ground zero are making our calculations more difficult, but I suspect the estimate is correct."

"Why is that, Major?"

"The yield is consistent with that of an old Soviet style backpack nuclear device. The Central Asian Confederation is known to have supplied backpack nukes to the Iranians."

"Are you saying the Iranians did this?"

"We cannot confirm that, sir, but that is our best guess."

"But my earlier briefings indicated that the situation around the Persian Gulf was rapidly being defused." The Vice-President looked over at the National Security Advisor.

"Don't look at me, Tom. Those are the reports I got as well. But given the situation, you never know what some extremist group might do." A Lance Corporal entered the room and handed the Major a report. He rapidly scanned it. "Sir, the latest report in from NORAD indicates that our satellites picked up radio transmissions from both Iranian and Saudi radar installations regarding an aircraft exiting Iranian air space approximately one hour before the blast. In addition, we picked up a

brief burst of radio activity around the site. It seems that minutes before the blast there was a fire fight between an undeter-
mined number of Iranian terrorists and a Saudi patrol."

"Jesus H. Christ! Have the Saudis done anything yet?"

"Not that we know of, sir. But with the evidence they have,
we are projecting a retaliatory strike by the Saudis. We may have
a full scale war going on by morning."

"Where is the President right now?"

"He was informed at his speech the same time you were and
is on his way to Air Force One. He should be landing at Andrews
within two hours."

"Any recommendations?"

"Tom, I think we need to talk to the Saudis and the Iranians
as soon as possible. We have got to try to defuse this situation.
We, the world, cannot afford a major war in the Middle East. Oil
prices will go through the roof!"

"Harland, you know as well as I do that this is a decision that
the President will have to make. Given the election, he is going
to have to handle this one directly. We can't afford to step on his
toes. It could cost him the election. Major, can we get in touch
with the President on Air Force One?"

"Yes, sir. I have a line opened. As soon as they are airborne,
we'll be in touch."

"Good. Major, get me General Talmedge at the Pentagon and
Scott Majors at CIA. I'd like them to work up briefings for us
when the President arrives. Anything to add, Harland?"

"We should apprise Gibson at State of the situation. He is the
chief of the Mid-East Desk and will have a great deal to contribute. I'll get my staff to call him in immediately."

"What about the Secretary?"

The National Security Advisor gave the Vice-President a wry
grin. "Fortunately, he's out of pocket. Fishing in Montana. "

The Vice-President nodded. Hand picked by Congressman
Gallager, the Secretary of State was his voice on the Cabinet, a
voice that neither the National Security Advisor nor the Vice-
President cared to listen to. "For now we wait. Major, keep us

posted on any additional information, but we won't release any official statements until we brief the President."

✛

Captain Milos Nicholaevich had piloted oil tankers through the Persian Gulf to Saudi, Kuwaiti, Iraqi and Iranian oil terminals for almost a decade. He was sitting on the bridge that morning smoking a pipe while his senior officer navigated the narrowest section of the Straights when several sonic booms shattered the ever present drumming of the engines. The captain looked out of his window to see several aircraft rapidly disappear to the east. Moments later he saw the telltale smoke trail of Iranian anti-aircraft missiles chasing after the jets.

What he did not notice was the launching of the Chinese made Silkworm III anti-ship missile by the Iranian shore batteries. Whether the Iranians knew they were shooting at a civilian craft or whether they mistakenly took the huge tanker to be an aircraft carrier, he would never know. The missile hit amidship, tearing the supertanker in two. The initial fireball shattered the windows of the bridge, blinding the captain and knocking him out of his chair. The oil in the two tanks nearest to the site of impact erupted into flames as it poured out of the gaping breach into the water.

Additional tanks ruptured from the shock of the initial blast, adding more oil to the burning slick. After nine minutes the stern of the craft submerged entirely below the surface. The ruptured tanks in the bow slowly filled with water, dumping their content of oil out into the waters of the Straights. Several hours later, the bow also fully submerged below the water. The Straights were, essentially, blocked for purposes of commercial shipping, not that any commercial shipping would be traveling through the Gulf for the duration of the conflict. One survivor of the disaster hung on tightly to a piece of floating debris hoping to be rescued before being choked by the thickening smoke of the oil fire or before drifting into the burning oil slick.

✦

General Rachmid Al Rashad rose from his morning prayers in the courtyard of the headquarters of the Revolutionary Guard as the distant sound of anti-aircraft batteries shattered the morning silence. He and the others in the courtyard were startled. They, the heads of the Revolutionary Guard, had heard nothing of this exercise. Who had given the order for the firing?

Rachmid grabbed his coat and threw it on as he strolled back into the building. A worried-looking captain ran right into him, almost knocking him down.

"What is the meaning of this, Captain!"

"I'm very sorry, General. But we've had reports of aircraft approaching the city rapidly from the west. Our radar locked onto them as they crossed the border, but several of our missile batteries were knocked out before they could launch. They had not anticipated a hostile attack."

"But who?"

"Please, General, let's get to the bunker. We can talk when we get there." General Rachmid Al Rashad began to run behind the captain. The sound of anti-aircraft fire and missile launchings grew rapidly closer. They could even hear it through the several-foot-thick concrete walls of the corridor leading down to the Revolutionary Guard's command bunker. They turned the final corner towards the bunker when an explosion knocked them from their feet. The lights in the corridor cut off and a dimmer glow of the emergency lights flicked on to replace them. They rose and ran the last few steps into the bunker through thickening dust.

General Rashad saw Colonel Rabii shouting at a technician. "What is going on here, Colonel?"

"General, we have just been hit by the Saudi Air Force. Our aircraft are up and engaging them now, but the strike was totally unexpected. We've lost several missile batteries. They are over Teheran now." Another explosion, although not as strong as the first, shook the bunker right as the colonel finished.

"I can tell that, you idiot! But why? What do our intelligence sources tell us?"

"Nothing. The Saudis, weak and decadent as they have become, were going to lie down and accept our disclaimer of the actions of the terrorists and our apology. They would not react that way." A lieutenant ran up to the Colonel and the General.

"General." The General turned and glared at the lieutenant. He was not used to being addressed directly by such low-level officers. "General, I think that you should listen to this." Before the General could respond to the lieutenant's insubordination, the lieutenant turned to the computer console behind him and switched a few buttons. The large screen in the center of the room flipped over to television. The reporter was speaking in English, but a translation appeared at the bottom of the screen.

"This is Bernard Hoffman reporting from CNN headquarters in Atlanta. We interrupt our regularly scheduled program to bring you this important development. We have just received confirmation that a nuclear device was detonated over the Saudi Arabian oil fields within the hour. These Landsat photographs taken shortly after the explosion show the remnants of the mushroom cloud dispersing in the upper atmosphere. The lower clouds you see are from the burning oil fields."

"Unconfirmed reports from Riyadh indicate that the device was triggered by Iranian sponsored terrorists. Sources indicate that Saudi officials have in their possession tapes of radio transmissions from a patrol that engaged the Iranians immediately before the blast. We have received no official comment from the U.S. Government, but the President left Houston unexpectedly. We are told he will be arriving in Washington within the hour."

"We now take you to a helicopter flying close to the site of . . ." The reporter put his hand to his ear. "We've just received a report from our station chief, Tom Berringer, in Teheran. Several bombs have exploded over the city. Anti-aircraft missiles and batteries are filling the skies. Tom, do you hear us . . ."

"Shut that thing off, Lieutenant, and go back to your station. But keep watching CNN. I'll expect a report from you later about what the western press reports." Glaring at the colonel. "Evidently their sources are better than our intelligence. Now, Colonel, get me the President as well as the Ayatollah Rafsangi. We must craft a response. Place defense forces on full alert, but do not launch an offensive at this time. Is that clear!"

"Yes, sir."

Speaking silently, almost to himself, "I only hope that Rafsangi can control the mullahs long enough to let us develop a military response."

✛

10:05 P.M. EST
MIDDLEBURG, VIRGINIA.

Jason was lying in bed reading a book when the phone rang. Sam put down her needlepoint and picked it up. Jason looked over at her. Sam frowned.

"Jason, it's Bob Ryan." Jason looked surprised.

"Yes, Bob. What is it?"

"Jason, are you watching TV?"

"No, reading."

"Well put on CNN or any of the networks. Watch for a few minutes and then come on in."

"What do you mean come on in, it's 10:00 at night."

"I know. Just turn on the TV. You'll see why. We need to work on a few models by morning."

Jason hung up the phone. "Sam, flip on the television."

"What is it, Jason? Why do you have to go in?" Sam unconsciously rubbed her large belly. She was only a few months away from delivery. She reached over and flipped on the TV.

It showed a picture of Air Force One arriving at Andrews Air

Force Base. The President exited the plane and walked towards his limousine. He paused and turned back to the press podium. "Mr. President, are the. . ." He held up his hand until the reporters quit asking questions. "Ladies and gentlemen, I have scheduled a press conference for 10:00 a.m. tomorrow to discuss the crisis. As you know, I have just arrived back in town from Houston and have not had an opportunity to review the situation with my staff. That is all for now." Despite the continuing barrage of questions, the President walked calmly and evenly towards his limousine.

The newscast flashed back to the newscenter. "For those of you just joining us, we repeat the top story. A nuclear device was detonated on top of one of the major Saudi oil fields several hours ago, igniting the field and destroying one of Saudi Arabia's largest refineries and oil pipelines. Iranian-backed terrorists are thought to be responsible for the blast, although nobody has claimed responsibility."

Jason turned off the television. He and Sam both sat there for a minute. He then jumped up and began to dress. "You know, Sam, this is exactly what we needed. It's terrible, but it's perfect. In fact," Jason paused as he pulled on his jeans. "it's too perfect."

"What do you mean?"

"Think about it. We needed a justification to construct solar power satellites. Now we've got one. It's too much of a coincidence."

"You're not saying that Geoff or the Vice-President could have ordered the action? It was terrorists. It had to be."

"I know. I know. But it's just too convenient. Even so, I've got to go in and run some new models based upon the continuation and escalation of the crisis in the Gulf. We need to take advantage of the situation and give the Vice-President the ammunition that he needs to steer the President in the right direction before our opposition regains their balance. To do that, I'll be spending the next several days at the Farm." Jason had finished dressing and leaned over and kissed Sam. He patted her tummy affectionately. "I'll miss both of you. If anything happens, call the number I gave you. If you want to, move over to your father's for a few days."

"I hate to do that, Jason, but I might. I'll let you know. Drive carefully." Jason gave Sam a hug and a kiss and then left. Upon arriving at the complex Jason went directly to his office on the first level. He found Bob Ryan sitting in his office with the television on. Bob stood as Jason entered the room.

"You made good time, Jason." Given how closely they had worked together for the past months, they had abandoned the pretext of not knowing each other.

"I know. We've got a lot to do. But before we start, there's something bothering me."

"Shoot."

"Okay. We've discussed my concerns about the organization before. Because we don't answer to anybody and because we hold the future in our hands, I feel that some of the team members feel that we are beyond the law. We will be making the decision as to who will live and who will die, who will stay on the planet and who will have a chance to survive on one of the colonies. It's not a decision that I want to make, and, frankly, even though I am the titular head of this organization, I doubt that the decision will be left in my hands."

"But whose hands will the decision be left in? We all know, even if we don't acknowledge it, that democracy will not survive this catastrophe, at least initially. But what will replace it? I, for one, feel that there is nothing wrong with a benevolent dictatorship. However, they never last. Will the committee survive to vote on issues or will one member of the committee usurp the authority at some point down the road? I don't know. But, at this time, I don't think that we have much choice. If we don't get the colonies built, my concerns are irrelevant."

"Spit it out, Jason. I've known you long enough to know when you are digressing. You don't need to preface your point with philosophical bullshit."

Jason took a deep breath. "Okay. I think that one of the members of the committee engineered the bombing in Saudi Arabia." Bob shook his head.

"Jason, you're overreacting. We got a lucky break. That's it. Some nutty terrorist just did us a big favor."

"I don't know, Bob. It just doesn't smell right."

"A mere coincidence. We've known that they've had access to nukes for years. It was just a matter of time. Fortunately for us, the time was now. Besides, who would do it?"

"I don't know." Jason shook his head. "Regardless of the cause, we have an opportunity. I've got to start working on it. If the Vice-President can convince the President of the need for quick, bold action, we'll have a greater chance of getting our program implemented. The world and the voting public will be in shock for a while looking at the pictures that the press will keep in front of them. The Saudis have retaliated. The ensuing war will keep the public's attention."

"I think you've got something, Jason. The President has always been a dove. We can give him an alternative to getting entangled in a foreign conflict and win him the election." Jason smiled.

CHAPTER 15

MAY 30, 2008.
7:58 P.M. EST.
WASHINGTON, D.C.

The House Chamber in the Capitol Building was filled to capacity in anticipation of the President's address on the Middle East Crisis. As usual, the President received lengthy applause as he entered the room.

He shook hands with a number of Senators and Congressmen on his way down to the podium, stopped for a minute to say a word to Congressman Gallager and shook hands with the Speaker of the House and the Vice-President. As he stepped up to the rostrum, he signaled with his hands for the applause to stop. It died away slowly. The President paused for a moment to let the silence sink in. He then began his address.

"Mr. Speaker, Mr. Vice-President, members of the House and Senate and fellow citizens of this great country, I come before you today to address the current global crisis. As all of you know, several days ago a small nuclear device was detonated in the desert of Saudi Arabia over a major oil field. The source of the explosion has not been definitively determined, and I am not here to speculate on the identity of the perpetrators of such a heinous action. The Saudis allege that the government of Iran sponsored the terrorist attack on their country and launched an immediate military response.

"While war has not been declared by either party, a fierce air battle is underway. I have already sent an official mission to both nations urging that the dispute be resolved peacefully and that restraint be exercised. However, the peace process will take some time. Both nations deny that they have any nuclear arms, but I have also urged that no nuclear devices be used again.

"Several oil tankers have been sunk, one in the Straights of Hormuz. As a result, shipping in the Gulf has come to a complete halt. As you may know, the Gulf nations still produce forty percent of the oil available for export around the globe, the vast majority of which is shipped through the Straights. Since the crisis began, oil prices have skyrocketed, rising over two hundred percent. Stock markets around the globe have plummeted. While experts are still debating how long the fires in the oil field will last and the long-term effect of the detonation on the oil reserves in the field, oil will not be flowing from the Gulf region for weeks, if not months.

"Tonight I am directing that the strategic reserves of the United States be tapped and sold to U.S. suppliers at pre-crisis prices to ameliorate the shock of the increase in oil prices on the domestic market. However, the reserve will not last forever. I have also implemented an arrangement with our Mexican neighbors to increase their oil production and their exports to the United States. Emergency tax credits will be given to U.S. oil companies to help increase the exploration and development of the Mexican oil industry.

"In addition, I have directed that several proven reserves which to date have not been exploited due to environmental concerns be opened for limited exploitation. I acknowledge my promise to you that I have made throughout my political career to protect the environment. I am not now backing down on that promise. Unfortunately, as a result of the current crisis, I have determined that it is in the best interest of the nation to temporarily open up those reserves. I assure all of you that we will take the utmost care in connection with exploration and exploitation of resources in environmentally sensitive areas.

"Finally, I am prepared to submit legislation granting tax credits for the exploration and use of alternative energy sources such as coal, gas, wind, geothermal and oil shale reserves. I have also issued an executive order to the Department of Energy to speed up the approval process for constructing new nuclear plants. The oil locked in oil shale in the western United States exceeds the total reserves of all of the Gulf oil producing states by several hundred times. The cost of exploiting such oil has always been prohibitive. But now is the time to make a commitment to explore and develop such reserves. Unfortunately, it will take sometime before we are able to effectively tap our oil shale reserves.

"Many companies have already developed energy efficient consumer devices and automobiles. I have directed that additional efforts be made to produce electrically powered automobiles. Many have been developed over the past decade and usage has greatly increased in states such as California. However, electrically powered automobiles still represent only five percent of the market. In order to accelerate the demand for such automobiles, I am directing that the federal government phase in a purchasing program to replace all federally owned, non-military gas powered vehicles to electrical and other alternative energy driven vehicles over the next decade.

"Unfortunately, even with all of the steps that I have highlighted tonight, we will still need to tighten our belts. I urge all of you to limit your use of gasoline to a minimum. The efforts of each and every one of you in cutting back will save more oil than is contained in the strategic reserves.

"Of all of the measures that I have outlined tonight, some will help immediately and some will take time. However, we will still be tied to oil and, hence, to the Middle East. Relatively stable oil prices have lulled us into a false sense of security. Oil remains the lifeblood of our economy and of the global economy. We and our children remain the hostages of the exporters of oil as the current crisis, and the reaction of financial markets around the world, reflects.

"Many people, including some of my top advisors, have suggested that we again send our children out to fight to protect our sources of oil. Well, I say to them: No! We have sent our children over to the Middle East countless times in the past. Enough is enough! The price of oil is too high when that price is paid in the blood of our children. I say tonight that I do not intend to and will not commit U.S. troops in any capacity to resolve the current crisis. We have done so too many times in the past to no avail.

"What I do propose is a bold new step, a step to free this great country from the shackles of an oil-based economy. We will no longer be held hostage to OPEC. The Middle East has frequently been classified as the most important strategic area for the United States outside the U.S. mainland. Well, I intend to end its strategic importance. Rather than invest uncounted lives in a conflict that may take months or years to resolve, I intend to free us from our dependence on oil.

"Tonight I am reminded of the bold steps that Americans have taken in the past: the construction of the continental railroad and our conquest of the continent; the building of the Panama Canal, linking the world's two greatest oceans; and the development of the interstate highway system, to name just a few. Americans have always boldly looked to the future. John F. Kennedy stood here in this chamber almost fifty years ago and declared that we would put a man on the moon in a decade. We did it.

"Tonight, I direct that we declare our energy independence by taking an equally bold step, a step that will make us the largest exporter of energy in the world by the end of this decade. We will achieve this goal. We will do it for a fraction of the cost of the Apollo program and for far less than the cost of the oil we import each and every year.

"I can see all of you asking: How will we do this? The answer is; the unlimited energy in space. We will build solar power satellites to collect that energy and send it back to the planet for use as electricity. The amazing thing is that it is so simple. Unlike the Apollo program, we need no new technological breakthroughs. We can do it with existing technology. The only thing lacking in the past was the commitment. And I am here to make that commitment. We will take the crisis that we presently face and turn it into a catalyst to reinvigorate America and boldly stride forward into the twenty-first century.

"By this time next year, we will have the first prototype satellite up and running. Within five years, we can and will replace all of the energy that we are currently importing. We can and we will do it, ladies and gentlemen. And it will not cost us the blood of our children.

"Thank you, God bless and good night."

CHAPTER 16

Australian officials have issued warnings for the coming spring and summer seasons to limit exposure to the sun. Satellite and aircraft observations have indicated that the ozone hole will be larger than usual. Officials fear that skin cancer rates will go up significantly if warnings are not heeded. Despite the warning, beach hotels are heavily booked due to the harsh winter anticipated for Japan and Russia. The locals, always skeptical, appear ready to flock to the beaches in droves despite the fifteen percent increase in melanoma-related deaths over the last decade.

11:15 A.M. EST.
CAPE CANAVERAL, KENNEDY SPACE CENTER, FLORIDA.

Beautiful warm weather enveloped the Cape for the launch of the Delta Clipper heavy lift vehicle. The Clipper would lift the first components of the prototype solar power satellite into orbit. A space shuttle with a team of trained astronauts already waited at the Space Station. Jason breathed in deeply. He could taste the salt in the breeze blowing off of the Atlantic. He stood next to Sam and her father on the VIP viewing stand. They all wore short sleeves in the 85-degree weather, a nice change from the cold drizzle they had left behind earlier that morning in D.C.

Jason had not been invited to the launch but had tagged along with his father-in-law. His activities, which had contributed so heavily to this moment, remained a closely held secret. But Mr. Whitlock's companies were key contractors for the construction of the solar power satellites. Given the advanced notice that they had received from Jason and his team, Whitlock Industries had already begun the development of several key components of a solar power satellite as well as the Earthbound receiving stations. As a result, the competition could not match Whitlock Industries' bids. In fact, Whitlock Industries had purchased several thousand acres of New Mexico desert and set up a research and testing facility before the President's famous speech.

The speech had drawn immense fire from both sides of the aisle as well as from the press. However, the first polls run after the speech showed that the President's action had broad-based support from the public. He won the nomination of his party on the first ballot and easily beat his Republican opponent, despite the signs of a deepening recession. In fact, the President used the recession to defend his policies; and, to the surprise of many political pundits, the argument worked.

Now, Jason could see both the President and the Vice-President at the forward viewing stand waiting for the launch. The gathering of VIPs at the launch was greater than for any launch since Apollo 11, an indication of the success of the President's program.

As a result, security was extremely tight. The President's course of action had not won any supporters in the Middle East. The Saudis had expected U.S. support in their action and were irate when it did not appear. The war with Iran raged on and while the Saudis had the technological edge, they were suffering heavy casualties. Without the support of the United States, they had resorted to hiring mercenaries from numerous developed nations and from Russia. A quick outcome was not expected. Both Iran and Saudi Arabia were bloodied, but neither could defeat the other. No ground war had ensued due to their lack of a common border, Iraq having managed to stay out of the fray.

But the air war continued, totally interdicting shipping. Both countries were hurting, as was the world economy, due to the reduction in the flow of oil. It looked as if peace would be reached eventually as a result of the inability to militarily resolve the crisis.

As Jack Watson had predicted several months before, both sides blamed the United States. It was the usual no-win situation; if the United States deployed troops to defend the Saudis, there would be a loss of U.S. lives as well as condemnation from many radical Arab states. The United States would again become the great power trying to exert colonialism. The Saudis would be lackeys of the United States. By not entering the conflict, the United States had abandoned an old ally.

If the President's proposal worked and, as one of the architects of the proposal, Jason was convinced that it would, the power of the Arab world would be greatly diminished. That was the real reason behind the abuse laid upon the United States for its stance. Hamilton Smith had briefed all of the members of the executive committee of Project Runaway that the CIA had already thwarted several terrorist attacks within the United States and that additional attacks were expected. Only two days before, the CIA had located two new French-made shoulder-mounted SAM missile launchers in Jacksonville. They had not caught the terrorists that had hidden them.

As the launch approached, silence rose within the viewing area. All eyes turned toward the many launch clocks. Jason saw the fire under the craft as the main engines ignited. Moments later a deafening roar reached the viewing stand. The Delta Clipper rose majestically upwards on its pillar of fire, rapidly ascending into the air.

✦

As launch time approached, several heads popped out of the sunroofs of the cars parked along A1A. Out of one dark green van, two heads popped up, dropped back down and then reappeared moments later with what the eyewitnesses thought at first was a high-powered telephoto lens. Just as the Delta Clipper's engines fired and the ship began its ascent, one man tapped the other on his shoulder and ducked down into the van. The other one squeezed the trigger of the SAM. A small supersonic anti-aircraft missile shot out of the front of the tube.

Those around the van that noticed the launch of the missile were shocked. It was headed straight for the Delta Clipper. The man ducked back into the van and the other reappeared with another similar looking tube. Before he could fire, a local police officer patrolling up and down A1A on horseback, recovering from his initial shock, jumped off of his horse, pulled out his service revolver and shot the man in the van.

The shot didn't kill the terrorist but struck him in the arm, causing him to launch the SAM down A1A rather than at the Delta Clipper. A large double-decker tour bus with an open top run by a local company as the "best way to view a launch" became the unintended victim of the missile. The SAM detonated in the bus, tearing it to shreds, spreading a rain of deadly shrapnel around the vehicle and igniting the cars parked around the bus.

The police officer kept his head and approached the van cautiously. He heard the scream of "Allah acba!" come from the van and then was knocked over by another blast. He survived, but twenty-three others did not. Over sixty people were injured. What investigators later determined was that the attack had always been intended as a suicide mission and that, after the launch of the second SAM, knowing they wouldn't have the opportunity to launch their third and last SAM, the terrorists exploded a bomb in their vehicle.

Fortunately for the two astronauts on board the Delta Clipper, the defensive perimeter around the Space Center worked. Defensive missile batteries locked onto the first SAM and shot it down well before it got anywhere near the spacecraft. Helicopters dispatched to the site of the SAM launch arrived moments after the terrorists had blown themselves up.

⟡

2:45 P.M. EST.
KENNEDY SPACE CENTER.
PRESS CENTER.

"And I repeat, we in the Congress fully support the President's stand. The President's bold action in declaring our energy independence will enable this, our great country, to continue as the leader of the free world. We must not, no, will not! let terrorists dictate our actions."

"Congressman Friedman, do you believe that we should strike back at the parties responsible for this actions?"

"It depends on what you mean by strike back, Barbara. Will the perpetrators of this heinous act be brought to justice? YES! Twenty-two citizens were killed by the cowardly acts of these terrorists. I urge the President to use all means possible to track down those responsible and bring them to justice. Actions like this can not be tolerated. The Congress will fully support the President in this matter. And we will not be deterred from our goal of energy independence. America will not be bullied. Thank you ladies and gentlemen. Now I've got to get back to Washington."

Congressman Friedman left the podium and went to his limousine. He entered the limo and sat back, making sure he was out of camera range before pouring himself a scotch. He took a draw, "Well, Hamilton, we couldn't have planned that better ourselves."

Hamilton Smith, seated across from the Congressman, responded. "You're right Congressman. I wish I could take credit."

"Who did it? The Saudis?"

"That's the best guess right now. But I doubt we'll ever prove it."

"We don't need to. Those Sand Niggers will get what's coming to them eventually. No reason to hurry things. Do you think they uncovered our role in the catalyst?"

"No way, sir. Michael Barnwell was good. The Mosad knows something, but they won't let on. Intelligence both here and abroad think it was a small rogue Iranian splinter group. The second most popular theory is that it was Iraq. However, there's no proof either way. Nobody thinks it was us."

"Good. Now, what about our friends at Project Runaway..."

CHAPTER 17

As was always the case on a holiday weekend, powerboats tow-
ing water skiers, fishermen, jet skis, wind surfers and sailboats
filled the lake. Peter and Elise had worked hard to save up for
their sailboat, a twenty-four footer, and, despite the crowd,
were determined to give it a thorough test. They spent the night
anchored in front of Lake Lanier Islands to watch the fireworks
and were now, after a long day of sailing, primarily occupied by
dodging other boaters, making their way back up to the north
end of the lake to leave.

Peter set the anchor and walked back aft to sit with Elise,
drink the beer she had waiting for him and watch the sunset.
They planned to spend the night on board again, dock the boat
tomorrow and return to Atlanta in the morning. Peter had just
taken the first swallow from his beer when he noticed a sudden
chill in the air. The chill surprised him given the new record
high of 109 set the day before. He took another pull off his beer.

"Peter?"

"Yes, dear?"

"Would you mind going below and getting my rain jacket? I've gotten a sudden chill." Peter looked up. There wasn't a cloud in the sky. He shrugged his shoulders, took one last swallow of beer and went below. It was the last swallow of beer he would ever have.

High above their heads, a peculiar new atmospheric phenomena was occurring. Given the rise in temperatures in the atmosphere, the temperatures between the different layers of the atmosphere began to vary. The barriers between separate layers were thinning. The colder layer above the warm blanket of air at the surface had begun to occasionally dip into the warm layer below it. The occurrence had yet to be noticed because it generally occurred over the three-quarters of the globe covered by the oceans and was very localized.

However, this time, it would be noticed. Cold air slowly poured down a channel created in the warm air cooling the temperature at the surface. As the cold air penetrated to the surface it opened a small hole in the warm air layer, allowing the cold air to suddenly flood down to the surface in the form of a massive microburst, commonly known as a wind shear.

Just as Peter went below deck, the wind shear hit with the force of a tornado. The 200-mile per hour downdraft hit squarely on the boat, shoving it below the surface, drowning both of the boat's occupants. One eyewitness sitting in the bar at the marina saw the boat suddenly disappear below the surface of the lake and noticed the appearance of a large wave. The wave expanded rapidly outward from the epicenter of the microburst, reaching a height of ten feet when it hit the marina. But the wind from the downdraft reached the marina first, capsizing most of the boats tied up at the dock, snapping pine trees in half and tearing the roof off of the marina.

The wave expanded outward from the epicenter, slowly losing force. Before it dissipated several unlucky boaters hit the wave and capsized. Investigators at the site at first thought that a tornado had struck, but eyewitnesses dismissed that possibility. The sky had been perfectly clear. The National Weather

Service announced that the event was an abnormally large and extremely rare, fair weather microburst.

✦

MIDDLEBURG, VIRGINIA.

As much as Jason hated to leave Sam and Jason, Jr. at home, he relished this trip, his first trip to the solar power satellite receiving station in New Mexico. The trip out from Dulles International proved to be relatively uneventful. The only memorable event happened when the flight attendant spilled a drink on a sleeping passenger across the aisle from Jason when some unexpected turbulence hit the plane. The flight attendant quickly apologized and the lady didn't make a scene, but the look on her face when the ice slipped down her blouse, waking her, had been rather funny.

After landing, Jason took the Santa Fe International Airport subway system to the heliport at the far end of the airport. He waited in the air-conditioned waiting room while the Whitlock Industries helicopter prepared for takeoff. Jason didn't officially rate his own helicopter, but he had scheduled his flight to land an hour before the daily flight to the Whitlock Industries Rectenna compound. While waiting, Jason picked up a copy of the local newspaper and began to read through it. He flipped through the paper, scanning articles rapidly. One story on page two was about a tragedy that occurred in Atlanta, Georgia. Evidently, a freak meteorological incident had sunk a sailboat, killing two people. Jason read on. The phenomena merited further investigation. He made a mental note to look into it upon his return.

A man in a blue jumpsuit interrupted Jason's thoughts by clearing his throat. "Mr. Graham."

"Yes."

"We're ready for you. Please follow me."

"Thank you." Jason picked up his one small bag and followed the man out the door and into a blast furnace. He had read the weather report before leaving Washington and had dressed in short pants and a polo shirt, but there was a difference between reading about 132 degrees and actually experiencing it. In the short walk from the waiting area to the helicopter, Jason worked up a healthy sweat. He licked his lips. One side dried before his tongue reached the other side.

Fortunately for him, the helicopter was air-conditioned. He sat back to enjoy the trip. After twenty minutes, they had left Santa Fe and its suburbs far behind. Jason looked out the window at the mountainous desert below him. Jason's fellow passenger pointed out the Los Alamos complex off to one side. They arrived at the several thousand acres within the Chaco Mesa region that Whitlock Industries had purchased. Jason could already see the transmission towers for the power lines running from the facility to Santa Fe and Albuquerque. Jason hoped that several thousand megawatts of power would surge through those lines into the southwestern power grids tomorrow.

Finally, Jason could make out the mesh of the antenna farm that constituted the collection point for the microwave transmission of the power down from the satellite. The rectenna, as the entire antenna farm was known, would collect the microwave transmissions, pass the energy to the conversion generator and convert the energy to electricity. Several of the Whitlock Industries scientists had wanted to opt for laser transmission rather than microwave transmission. But, despite the advantages of a smaller collection field, too many members of the public, as well as the international community, objected to the thought of massive lasers orbiting above their heads.

Obtaining permits for microwave transmission had been a painful process in and of itself. Jason remembered the months of delay the initial test of the small prototype suffered as a result of lawsuits filed by citizens groups. Even with the full support of the Administration, it took months for the appeals to run their course and be defeated.

The press continued to print horror stories about what would happen to people, animals and birds if the microwaves drifted off of the target area. Jason remembered the picture shown in one tabloid of a poor bird that had exploded after being 'subjected to microwave energy'. The photographer must have stuck the bird in a microwave oven to achieve the result. It was ghastly. The headline said, "This could be you!" Finally the results of studies and the fail-safe mechanisms built into the system convinced most people not to fear it. Jack Watson later confirmed that the citizens' environmental groups were being bankrolled by a joint consortium of oil companies and auto manufacturers that stood to lose a great deal of business if the venture succeeded.

The stories had been particularly distressing to Jason, because he had worked on the software that would shut the transmission from the satellite down if it drifted off of the rectennas. In addition, the microwaves were very weak. It would take weeks of constant exposure to harm anyone. But the frustration had finally ended.

The helipad, located next to a concrete blockhouse, was at least two miles away from the outer edge of the rectenna. Jason followed his fellow passenger into the blockhouse. It contained a small room with a guard and an elevator. The elevator led to the belowground heart of the complex. A guard led Jason into a control room. It reminded Jason of the control rooms at the Kennedy Space Center. He was introduced to Dr. Gordon Templeton.

"Mr. Graham, it is a pleasure to meet you. I've reviewed your work. It's excellent."

"Thank you, Dr. Templeton. But, please, call me Jason. And it's my pleasure to meet you. You've made a reality of something that I feared I would never see."

"Enough. Let's just say that we both work hard for what we believe in. Now, Jason, let me show you around. Oh, by the way, my friends call me Gordon."

"Okay Gordon. But I'm here to work. You don't have to take the time to show me around."

"It's my pleasure. I'm just a figurehead here. Look around the room. Does it look like I will be missed?" Jason took a minute to look around at the technicians busily working away. "Things will continue to run smoothly if I leave for a little while. Besides, I rarely get the opportunity to show this place off to somebody that I know will appreciate it."

"But what about the politicians and citizen's groups that come to see the complex?"

"I give them the standard spiel in the auditorium at the public end of the complex and then turn them over to our PR specialists."

"I was wondering how you handle the press here, after all of the bad publicity this complex received."

"We try not to show them the control room if possible. Instead we take them to the above-ground backup control room at the public end of the complex. You read the press we got. How do you think the press would react if they found out that most of our facilities are located below ground?"

"They'd probably write that you had constructed the control complex below ground because of your fear of exposure to microwave radiation."

"I know. We built below ground primarily for security purposes, but I'm sure the press would ignore that. Besides, being below ground enables us to get directly below sections of the rectenna." They exited the control room and entered a service tunnel filled with cables and pipes. Dr. Templeton sat down in the driver's side of a golf cart and signaled Jason to get in on the other side.

"We've got over twenty miles of tunnels built already. I'll take you below the antenna field." They rode for ten minutes down an endless number of seemingly identical tunnels.

Jason didn't pay much attention until he realized that he'd never find his way back if something happened to Dr. Templeton.

"How do you keep from getting lost down here?"

Dr. Templeton laughed. "Oh, I could find my way around down here blindfolded, but the corridors and halls are all marked and numbered in case somebody gets lost. Besides, that badge the guard gave you has a miniature transmitter in it. Security keeps track of all personnel, visitors in particular. In addition, there are microphones at all tunnel junctions. You could call for help and we could either direct you out or send somebody to get you."

"That's good to know. By the way, given the number of miles you have to cover, doesn't it take a long time to get to, say, the visitor center?"

"Just over twenty minutes, but we're working on that. In fact, that's one of the things I wanted to show you." Dr. Templeton brought the cart to a halt. "I received a message from the CEO of Whitlock Industries. She requested that I show you our special project. I don't get the chance to show it off very much, so I'm glad to have a chance to do it." Jason smiled at the mention of the CEO of Whitlock Industries. Evidently, Dr. Templeton didn't know of Jason's relation to the CEO.

Dr. Templeton pushed several buttons on a keypad next to a large metal door. Jason detected a slight hissing noise as the door slid open. When they stepped inside the door, it closed behind them. They were in a small chamber with a similar door on the opposite side. While Dr. Templeton pressed several numbers on a similar keypad, he explained to Jason: "These doors are airtight and are designed to maintain a vacuum on the other side." The second door swung open and they entered a tunnel about the same size as those they'd been travelling in. Jason immediately noticed the large metal coils located along the floor and the ceiling of the tunnel. He smiled, guessing what Sam's surprise for him was.

"I don't know how familiar you are with magnetic levitation or superconductivity, Jason, but we've been given the task of developing a maglev train. You've probably seen the high-speed trains the Japanese developed to replace their bullet trains. The train is driven by magnetic repulsion created by electrical currents flowing through magnets along the 'track.' That train has been run successfully for the past several years at just under supersonic speeds.

"We are attempting to take that same process one step farther. We have tunneled one line from one end of the complex to the other. Once the tunnel is completed, the tunnel will be sealed and we will lower the pressure in the tunnel to the equivalent of, say the atmospheric pressure on the surface of Mars. It will be fairly close to a vacuum. This train will accelerate at one half a gee and will traverse the entire complex in just under four minutes."

Jason smiled. "That's incredible. But isn't it overkill for a complex of this size?"

"This is just a test facility. Eventually, this tunnel will run from this site to the site of the next rectenna field in Arizona. If the tests work, we'll crisscross the nation with an underground supersonic rail system. With cheap electricity from solar power receiving stations and the new tunneling techniques that we are using, there will be very little to keep the underground rail system from capturing a large percentage of the transportation industry."

"Did you develop the magnetic levitation system here?"

"Oh, no. Whitlock Industries designed the basic system years ago for use as a mass driver to accelerate raw materials off of the surface of the Moon. If they can successfully deploy a mass driver on the Moon, they should be able to reduce the cost of construction of solar power satellites by at least fifty percent. They decided to build this test track here because of the readily available power we'll have beginning tomorrow, when the satellite comes on line."

Jason pointed to the cables hanging down from the ceiling. "This tunnel doesn't appear to be ready. When is the first train going to be ready for its first run?"

"Tomorrow, in the main tunnel. You thought that this was the main tunnel? No, the main tunnel is currently being evacuated for the final test of the seals. I can't show it to you today, but hopefully we'll be going for a ride on it tomorrow for the press conference at the Visitor Center. For the time being, this will serve as a service tunnel. Ultimately, it will become a part of the tunnel from this complex to the Arizona complex."

"Let's keep going so we can finish the tour and get to work. I'm planning to finish the tour in the computer center. I'm sure you'll be interested to see it." Dr. Templeton patted Jason on the back and then led the way back out to the maintenance tunnels and onto the golf cart.

"By the way, I've been meaning to ask you, that is, if you don't mind; how do you rate the security clearance to see everything? You are just about the first person that's gotten to see everything."

Jason laughed. "I don't mind telling you, but I'd rather that you keep it as our secret. I don't want the other staff members giving me any special favors or treating me differently." Dr. Templeton looked at him quizzically. "You mentioned that you had received a directive from the CEO of Whitlock Industries. Well, the CEO just happens to be my wife."

"You're kidding."

"No, not at all."

"Lucky you." Dr. Templeton remained silent for several minutes as he continued to drive towards the rectenna field. "You know, if I hadn't looked over your work personally and if I hadn't heard rave reviews from our own software genius, I'd regret having given you the personal tour. But from your work, I know that you have earned the honor."

"Well, I appreciate that. And I appreciate the tour. What you've done here in a very short period of time is both impressive and important." Jason thought to himself that important was an understatement. So much of the plans to effectuate Project Runaway rested on the success of the solar power satellites. And there was so much that could have gone wrong and could still go wrong. Assuming that the satellite did work as designed and that several dozen megawatts of power were added to the grid, the next step would be to start building the cities in space, ostensibly to house the SPS work force.

When they returned to the computer center, Jason got to work checking on the software that he and his team had helped develop. He checked his simulations one last time. The software worked perfectly. Jason hoped that the real thing would work as well.

Jason stood behind the chief programmer as the transmission came on line. The sensors on the rectenna detected the microwaves and began converting them into electrical energy and adding them to the southwest power grid. Slowly, the power increased to twenty, thirty, fifty, ninety percent. Everything worked perfectly. A cheer went up from the control room as the transmission reached one hundred percent. Five gigawatts of power flowed into the power grid, providing more energy than the largest oil-fired plant in existence and more than all but the largest nuclear plants.

Readings indicated that power at the center of the rectenna was within the design parameters and that no leakage had been detected around the perimeter of the rectenna field. Jason breathed a sigh of relief. Dr. Templeton invited him to the press conference, but Jason declined the invitation. The executive committee had decided that Jason and the others working at the Farm should keep as low a public profile as possible. While Hamilton Smith had assured them that he had a handle on security, they didn't want to give some hungry reporter an angle to begin an investigation.

Jason continued to spot-check various elements of the software and it continued to check out. He was just about to shut down for a late lunch when a bell rang out over the intercom. All heads in the room turned. Jason leaned over to the programmer sitting next to him. "What's going on?"

"I don't know. That bell sounds for emergency announcements. I have only heard it in tests before." Jason began searching the screens in front of him for an indication of the problem. He saw nothing. He stopped and listened when Dr. Templeton's voice began over the intercom.

"Ladies and gentlemen, this is Dr. Gordon Templeton, project manager for this complex. I have an unfortunate announcement to make. We have just received word that President George Wright is dead!" While the room had been silent when the announcement had begun, the silence that followed the announcement was deafening. "The President was on his way to the complex when Marine One crashed in the mountains between here and Santa Fe. The cause of the crash has yet to be determined. Please remain at your stations and keep working. The Secret Service has requested that nobody leave the complex."

Jason, as well as everybody in the room, was stunned. After a few minutes of silence, he heard the whisperings of several people over the hum of the equipment. He went to call the Farm but thought better of it. Given that the President had just been killed, there was too much chance that the telephone lines would be monitored. Instead, he called Sam at her office. After a short pause, Sam's secretary picked up the telephone. She told Jason that Sam was in her father's office watching the news reports. She transferred the call.

"Jason, how are you?" Sam's voice broke up a bit, as if she'd been crying.

"A bit numb, but okay. How are you? You sound shaken."

"I'm all right. You know, there's something about the death of a President, it gets to you."

"I know."

"Even though I didn't vote for him, I can't help but mourn. I feel as if I've lost something, some part of myself. My father was just telling me that he felt the same way when President Kennedy was shot years ago. Even though we may not have agreed with President Wright about many things, he was a symbol of our nation."

"I feel the same way. Even more so now, as a result of his support of the project. By the way, I haven't heard anything out here yet. I don't know the lay of the land and haven't found a television. Have you heard any details? Was the helicopter shot down or sabotaged?"

"Nothing official has come out yet, Jason. The press indicated that the weather was clear, but there's no evidence of foul play. The Marines and the Secret Service haven't released any information yet. But the press is speculating. Several press helicopters were turned away from the scene by Air Force aircraft."

"Have they said anything about the President's mission yet?"

"Several of the networks did mention that he was traveling out to the site to commemorate the coming on line of the first solar power satellite. But nothing much has been said about the plant yet. You don't think that they'll shut you down; do you?"

"No! At least, I don't think so. Why would they? Everything's working perfectly. It's too bad that the President didn't live to see the fulfillment of his vision." Jason paused. "I doubt I'll be able to catch a flight back tonight. I'll get back as soon as I can. Love you sweetie."

"I love you, too. Goodbye."

By the time that Jason got back to the control center, the large status screen in the center of the room had been converted to a television. All eyes in the room focused on the report of the crash. The scene switched from a distant view of rising smoke to a reporter standing in the midst of a multitude of shouting, jostling reporters at the crisis center set up at the Santa Fe international airport. The reporter was trying to say something over the crowd when the camera refocused on a colonel at the podium.

"Ladies and gentlemen, can I have your attention? In order to quash any speculation about the cause of the crash and the death of President Wright, I'm here to give you a brief report. At 2:04 p.m. Mountain Standard time, the President took off from this airport in Marine One for an unscheduled appearance at the Chaco Mesa power station. At 2:18 p.m., the pilot indicated that they had encountered a mild downdraft. Seconds later, the pilot radioed a mayday and radio transmission was cut off. One of the two escort helicopters also crashed. The remaining aircraft managed to set down next to the crashed aircraft without incident. All personnel aboard Marine One were dead. The copilot of the other helicopter survived but is in serious condition and is being transported to the hospital.

The pilot of the undamaged helicopter reported that Marine One was literally knocked out of the sky. Seconds later, a severe updraft buffeted his helicopter. The evidence and eyewitness reports indicate that an unusually strong wind shear hit the President's helicopter without warning, causing the crash. A more detailed report will be released as soon as the National

Transportation Safety Board investigation team concludes their study of the incident. That is all."

The colonel turned and left the podium despite the barrage of questions that followed him. Jason thought back to his flight over the same area earlier in the day. Something was nagging at the back of his head. Of course! The article that he'd read in the newspaper at the airport. Those two boaters in Georgia had died as a result of a similar incident. Jason had no doubt that both incidents were related. He was certain they had something to do with global warming.

CHAPTER 18

Norwegian fishing fleets have ventured farther north in their quest for a full catch than in previous years. The temperature and salinity of the water seem to indicate that more ice than usual is melting from the polar cap this summer, again raising speculation about global warming. Several scientists have already reported that the ice cap is diminishing in size and that the record temperatures in northern Scandinavia are an indication of a serious problem.

However, Nobel Prize laureate and noted environmentalist, Dr. Thornton Hildebrand, disputes the claims of global warming. He points to an increase in sunspot activities and the decline in the ozone layer above the Arctic as the cause of the melt-off this year. He predicts that it will be a temporary phenomena.

Nonetheless, the Norwegian fisheries foundation has commissioned a team to study the condition of the Arctic over the next several years to determine the effect of the increase in the seasonal melting of the ice pack on migratory fish.

✦

CAMP DAVID, MARYLAND.

Jason had never been to Camp David before. But now, with Tom Hammond the President, he figured he'd be seeing a lot of it. The press had always respected the privacy of Presidents at Camp David. Nevertheless, to insure privacy, all of the members of the executive committee were flown in.

A marine corporal led Jason to the elevator of one of the many underground bunkers located at Camp David. To his surprise, Jason bumped into General Maxwell in the elevator. This would be the General's first meeting in some time. Jason had just finished downloading his data into the computer when the President entered the room. Everybody stood up.

"Ladies, gentlemen, please, sit down. We've been meeting for several years. There's no need for formality here." Everybody took their seats. "Before we begin, let me briefly express my sorrow at the passing of George Wright. While many of you may look upon the event as fortuitous," the President focused his gaze on Hamilton Smith and then General Maxwell, "we were fortunate that George was the President at this critical moment in history. We all owe a great deal to him and his ability to persuade the public to back the development of solar power satellites. I doubt that I could have accomplished what he did."

"Now, without meaning to sound callous, its time for us to take advantage of the opportunity that fortune has presented to us and use the death of George Wright as a vehicle to keep the solar power satellite program growing. I will travel to the Chaco Mesa station later today to rename it the George Wright power station. In addition, I intend to announce the construction of the George Wright Space Station and the establishment of a lunar base to enhance the construction of more solar power satellites."

"The forces of the opposition, spearheaded by Congressman Gallager, are martialing even now to eliminate the solar power satellite program. While they haven't begun to publicly oppose the program yet, the handwriting is on the wall. Congressman

Gallager and his backers have opposed me from day one. Were it not for George's personal loyalty to me, I might have been dropped from the party ticket."

"We can't afford to underestimate Congressman Gallager. He has more support from my own party than I do. Granted, they can't publicly denounce the solar power satellite system yet. The public isn't ready to see anybody, even indirectly, besmirch the name and the works of a dead President. But, as you all know, the voting public can be manipulated, especially when it comes to their wallets. As a result of the peace reached in the Middle East, brought about in part by the success of the first solar power satellite, we have learned that OPEC intends to lower the price of oil to a point where the cost of building solar power satellites will not be justified. I know, I know. They can't keep prices that low for very long. But they can do it long enough to turn public's sentiment around."

"We have got to move now and move quickly, while the public is still on our side. The new Space Station and the lunar base, if they come on line quickly, should get us beyond the point where Gallager and the other luddites in Congress will be able to stop us. They will, doubtlessly oppose my move, but I have the votes right now. They know it. They can't beat me now. But they also know that I will not have the votes six months from now."

"So let's keep that in mind as we move forward. One thing they cannot be counting on is that we have already done a great deal of work on the Moon. We can deliver everything ahead of schedule and under budget. That should raise some eyebrows and help our cause immensely."

"One more thing before we get started; while this Program is double-black, that doesn't mean that I am willing to give the members of the team license to do whatever they deem is appropriate for the good of the Project." The new President glared at Hamilton Smith. "Now that I am President, I plan to use my office to benefit the Project; but I also plan to keep a firm grip on things. I will not tolerate covert actions being undertaken without my prior consent. Granted, some actions will need to be taken that, under public scrutiny, couldn't be taken. That is the purpose of black and double-black programs. But keep in mind we are still answerable to the people and to the future. Despite our backs being, literally, up against the wall on this one, the ends do not necessarily justify any means."

"What was done in Saudi Arabia needed to be done. However, a few extra days would not have made a difference. I will not have similar actions taken without my knowledge! I know the argument that the Commander-in-Chief needs to be insulated from such an action. In this instance, I don't give a damn about the consequences. If this breaks before we are prepared, we'll all be up the creek. Now that I've made myself clear, let's get on with the meeting." While Jason was relieved by what the President had said, others were not. Hamilton Smith hadn't even flinched, but beads of sweat had formed on General Maxwell's brow. Jason wondered what he was up to.

Jason didn't take the time to think about it. As head of the Project, it was his job to run the meeting. They had a great deal to cover. "Thank you, Mr. President. Before I start, there is one bit of data relating to the death of President Wright that has not been made public. After studying the accident, our team has concluded that unusual and previously unknown atmospheric anomalies are beginning to develop. The crash was the result of unusually large and powerful microbursts. Microbursts, which are also known as wind shears, have been known for years as a cause of aircraft accidents. Fortunately, with the developments and deployment of sophisticated Doppler radar at most domestic and international airports, no crashes have resulted from microbursts for years."

"That is, until now. But these are a new type of microburst. They are not in any way associated with thunderstorms and can strike in clear weather. Due to the change in temperature of various levels of the atmosphere, cold air is occasionally bursting

through the hotter levels below it in a rapid plunge to the surface. The forces exerted can easily be in the range of several hundred miles per hour."

"Our computer models predict that while these events are
currently very rare, they will occur with increasing frequency
over the next decade as the atmosphere adjusts to the increased
temperatures."

"Unfortunately, while we can set up a warning network to
avoid future air crashes, doing so could prove disastrous. The
revelation would doubtlessly lead to an investigation of the
cause of the atmospheric instability, which could lead to the
public learning of the impending ecological disaster." Jason
turned slightly to address the President directly. "Mr. President,
I propose that we withhold our knowledge of the cause of your
predecessor's death from the public."

Most of the heads in the room seemed to nod in assent with
Jason's recommendation. "I concur. However, Jason, as
President, I will be traveling frequently by air. Are you telling
me that I, that all of us, will be subjecting ourselves to a heightened risk in flying?"

"To some extent, yes. However, the events will remain
rather rare over the next couple of years. It was extremely
unlucky that your predecessor happened to be where he was
when the microburst hit his aircraft. Over the next couple of
years, you are more likely to be struck by lightning than to be
the victim of one of these microbursts. In fact the odds of
being killed by a microburst are only slightly better than winning the national lottery."

"For the time being, I wouldn't worry about it. But I do propose that we develop a detection system to predict where a
microburst is likely to strike. That way we will be able to notify
members of the committee, as well as other key personnel, not
to fly in a particular area if a microburst is predicted. In the long
run, air travel will become very hazardous. But by then, we
should have an underground magnetic rail system in place running to and from our key surface facilities."

Geoff Barnes interrupted. "But, Jason, I know for a fact that the National Transportation Safety Board is conducting an extremely thorough investigation of the President's crash. Won't they put two and two together?"

"That's one of the questions that we considered when we first got the results from the computer. It is possible, but we consider it to be unlikely. The new microbursts are too rare right now to be properly studied. And the investigators do not have access to either our software or to our database. Without them, the cause of the crash should be written off as a freak occurrence."

"Now, let's get back to the agenda."

✦

General Maxwell wiped his brow with a handkerchief as he stepped into his staff car. He hadn't liked the summons to Camp David or President Hammond taking a firm grasp of Project Runaway. Graham was a little snot. He could be controlled. Besides, according to Major Ryan, Graham had no clue what was going on behind the scene. A sitting President would be a different matter. General Maxwell lifted the phone and called Congressman Joel Friedman.

"Joel, Hammond came in full of piss and vinegar. If we do anything, he'll find out."

"Relax George. We don't have to do anything right now. You've already taken care of the creation of the Black Berets. If Hammond even learns about them, he'll be told that they are an elite inter-service unit developed to respond to terrorist activities. He doesn't have to know where their loyalties lie. Hammonds has been a dove too long. He's not popular in the Pentagon."

"But he is the President."

"And so was George Wright. He didn't know shit. Don't worry George, Hammonds will be a one term President. When Barnes runs, he'll need to turn to me for support. Relax. I've got everything under control."

CHAPTER 19

Several hundred thousand are reported dead in India and Bangladesh as a result of fierce storms and flooding. Thousands of bodies have washed into rivers and into the Bay of Bengal. The loss of housing and the threat of disease from the dead and decaying bodies could result in the deaths of hundreds of thousands more. The government of India is pleading for international assistance.

Kalinin Village, Russia.

Shots rang out and smoke filled the air in front of the muzzles of the two shotguns. Two grouse fell to the ground.

"Great shot, Nicholai Stipanovich!"

"Thank you, Victor Borisovich. I do believe that I got them both." Nicholai Stipanovich Greschko smiled to himself as the dog trainer ran out to pick up the fallen birds. He was pleased to have gotten a double, a feat rarely accomplished despite his prowess with a shotgun.

"No doubt, I shot behind the birds."

They reloaded and continued to walk through the fields that surrounded Greschko's dacha. "You know, Nicholai Stipanovich, I have been worried about the money being spent on your effort to build a base on Mars. You have pulled funds away from the production of aircraft that we need for the defense of our country."

"But who will attack us, Victor Borisovich? The United States? Bah! Germany? No, we do not need to live in fear of attack any longer."

Kirov halted in the middle of the field. "I never thought I would hear such words from you, Nicholai Stipanovich! You built your company around the backbone of defense spending."

"You are right, of course, and you will never hear such words from me in public. But between us, admit it. I am correct."

"With regard to Europe and America, yes. But what about the Chinese or the Moslem nations?"

"They don't have the balls to do it. Despite our reduction in arms, we still remain the number two military power on the planet. They would not dare to attack."

"You may be right, but why Mars? It's a dead end. You could make more money fanning the flames of the yellow peril. I could get you an order for at least another 50 MIG 38s and a couple hundred T97 tanks. If we work it right, foreign orders would double or triple that. And I could keep our defense forces more prepared." It was Greschko's turn to stop walking.

"Victor, we have known each other for years; have we not?"

"Yes."

"And we have each gained from the other?"

"That is true."

"I am asking you to trust me. I cannot reveal my reasons, but they are important and they are not motivated merely by profit. Support me in this one and we will both benefit. Trust me."

"But what is this about? Why Mars?"

"Just trust me. It would be unfortunate if you did not. This is something that we cannot afford to lose on." Greschko paused. He looked over at Kirov and then patted his gun. "I am reminded of the unfortunate hunting accident suffered by Konstantine Golganov."

Kirov stared into Greschko's eyes. Greschko knew he was not a man to be trifled with, but Greschko had his measure. Greschko did not blink. Golganov had been the Minister of Science and Technology before he had died several months before of a hunting accident at Greschko's dacha.

"Okay, Nicholai. You have my support. But you had damn well better be right. And if the pressure builds, I'll expect an answer from you."

"That you will have, Victor, as soon as I can give it to you."

Just then the flutter of wings disturbed their conversation. They both raised their guns to their shoulders. This time, however, Kirov got the double. He lowered the gun and smiled broadly at Greschko.

✦

BEIJING, CHINA.

General Hong Liu sat behind his desk in the Forbidden City studying intelligence reports when Minister Ping walked into his office unannounced.

"Have you found out what the damn Americans and Russians are up to yet!" Minister Ping paced in front of General Liu's desk.

"Calm down. No, I have not. My sources do not tell me why they have persisted in their new space ventures. We had expected and predicted the initial reaction and the new funding by the United States, but I have no idea why they have persisted in pushing back into space. All indications were that they would have abandoned their space station by now, not planned to construct far larger ones."

"It is destroying our propaganda victory at home. Things are unstable. This could hurt both of us severely."

"I know. But there is nothing we can do that we are not

already doing. Our sources in Washington are trying to uncover the motivation behind America's move but have been unsuccessful. We have been lobbying heavily both directly and indirectly. And our sources indicate that the indirect support and pressure that we are exerting on Congressman Gallager is working. It is unlikely that President Hammond will win his own party's nomination, let alone the Presidency. Once Gallager, or one of his confederates, is in office, our worries will be over. The United States will once again abandon its efforts and retreat behind its two moats. Their solar power satellite will become a dead end showpiece, like Apollo. We will be left to profit from their shortsightedness."

"I certainly hope that you are right. But what about the Russians?"

"It is clear to us that Greschko is behind the Russian push. His motive is profit. He is a greedy pig. I do not fear him. Let the Russians have Mars. What will they have gained? I doubt that even Greschko will be able to sustain the drive to establish a permanent presence. He will merely burn out the resolve of the Russian people. He will not be able to strong-arm the government forever. Besides, if you have looked at the reports on Russian military preparedness, you would see that his maneuver is hurting the military. Their troop readiness along our border is at an all-time low. Let him play his game. He cannot win."

"If you are wrong, we will both lose. Our people need to see success. The United States and Russia are both belittling our efforts."

"Let them. History is on our side. We have the patience to win. And we will. Our people will not rebel. The economy is still improving. We have not reached the crisis point."

CHAPTER 20

The Cousteau Institute reports that plankton counts in the Mediterranean have declined more than 90 percent over the past decade. After a strong comeback during the late '80's and early '90's, the Mediterranean is again on the verge of becoming a dead sea. While scientists originally attributed the decline to over 2000 years of pollution, they now suspect that the past decade's 3-degree increase in temperatures was the real culprit.

The Central Atlantic.

Numerous storms had formed that year over the waters of the Central Atlantic. However, the water remained unusually warm. A low-pressure cell drifted off of the coast of Africa and stalled over the warm waters. Immediately it began to organize and gain strength, pulling moisture and energy out of the warm waters.

The first satellite pass over the area noted the development of a tropical depression, but national attention was focused on

Lois, a Category 2 storm currently battering the outer banks of North Carolina. By the time the new storm began to move towards the west, it had already developed a small eye and had intensified into a tropical storm. In accordance with the list of names promulgated at the start of each hurricane season, the storm was named Marla.

When the storm first took aim on the Caribbean, the National Hurricane Center dispatched storm chaser aircraft. By the time the aircraft entered the storm, the pressure within the eye had already dropped significantly and sustained winds had picked up to over ninety miles per hour. The national media began to cover the progress of the new storm as the Gulf Stream, combined with a strong high-pressure cell, pushed Lois rapidly northwards, out to sea.

The pressure within Marla continued to drop as she approached the warm waters of the Caribbean. The storm stalled just beyond the Virgin Islands, brushing them with gale force winds but not unleashing the fury of the hundred and twenty mile per hour winds that swirled around her eye. She did not move appreciably for twenty-eight hours, but continued to gain strength. Reporters were already comparing her to the most powerful storms to ever hit the continental United States as sustained winds increased to 175 miles per hour.

Finally, as if having finally made up her mind, the storm turned and began to move rapidly towards the northwest. Unlike most fast moving storms, Marla continued to gain strength and size. Gale force winds already extended out thirty miles from the eye and tropical storm winds extended out well over one hundred miles from the eye. Given the present track, forecasters at the National Hurricane Center in Coral Gables projected that Marla would plow into the northern coast of Florida or the southern coast of Georgia.

FEMA began to implement evacuation plans, given Marla's ferocity. Her progress remained relentless and she continued to head for the coastline of southern Georgia. Shipping had been rerouted up and down the Atlantic seaboard and preparations for an emergency launch of the Yeager heavy lift shuttle were being made at the Cape. NASA officials feared that, given its strength, Marla might cause serious damage to the Yeager.

Workers aboard the Space Station Freedom couldn't help but look down at the massive storm as it neared the U.S. mainland. From space, the storm appeared to be as large as the entire Gulf of Mexico, with amoebic tendrils of clouds whipping off of the tight spiral around her eye. Marla looked poised to swallow the entire State of Florida.

Twenty-six hours before projected landfall, a hurricane watcher aircraft flew into Marla's eye. The officials at the National Hurricane Center listened incredulously at the readings radioed back from the plane. Pressure had fallen below that of Hurricane Gilbert, the former record holder that had decimated Cancun in the 1980s. The readings were by far the lowest ever recorded in the Atlantic. Hurricane Center officials estimated unheard of sustained winds of two hundred and thirty to two hundred and forty miles per hour.

In addition, the storm had turned toward the south, something hurricanes rarely did this close to land. Marla had taken aim at South Florida. South Florida was not ready. Officials rapidly issued evacuation orders for the entire coastline of Florida from Cocoa Beach south to Miami and back north to Tampa. Both coastlines of Florida braced for the worst. Given the strength and power that Marla displayed, she would, in all likelihood, still remain a category five hurricane when she passed over Florida and entered the Gulf of Mexico.

NASA officials worked continuously and took the risk, with the consent of the two astronauts that volunteered to fly the mission, of launching the Yeager in winds of over thirty knots. The alternative was to lose the Yeager to the storm. After being buffeted severely after launch, the Yeager arced over Marla, giving both astronauts a view of the storm as it began to come inland over Palm Beach. They were both glad to be where they were rather than in the path of the storm.

Even before the eye reached land, the sea had begun its

devastation. The waves destroyed the artificial reef constructed to defend the coastline and the rebuilt beach that had cost residents millions.

Winds the strength of an F5 tornado, but a hundred miles in diameter, punched through boarded up doors and windows, flattening the houses in its path. The thirty-five foot high tidal surge rushed ashore with the eye, sweeping away the debris created by the winds. The steel and glass towers of West Palm Beach fared no better. The winds blasted windows out, spraying shards of glass for miles in every direction. And while the winds did manage to topple a few of the older towers, it was not the wind, but the water that did the real damage. The massive tidal surge ripped foundations out from under buildings. The towers, constructed to withstand hurricanes of lesser strength, collapsed one by one to be swept away.

Fortunately, the storm did not linger. It passed rapidly over Florida and entered the Gulf of Mexico to continue on its trek of devastation. In the aftermath, pictures showed what was left of Palm Beach, but there was very little to see. The eye had passed directly over the art museum located on Lar-A-Margot, the old Margaret Meriweather Post estate. As helicopters flew over the site, nothing of the stately old mansion remained. Instead, the storm had carved a mile-wide channel where the house, and dozens of others, had stood.

The hurricane had driven the ocean well inland through the heart of Palm Beach, creating a new harbor. Destruction was near 100% from Fort Pierce to the north end of Fort Lauderdale. Extensive damage had occurred as far south as Homestead and as far north as Titusville. Even Orlando had suffered moderate damage from the storm.

Marla left Florida as the most devastating natural disaster in the modern history of the United States. It was the first time in over fifty years that more than a thousand people lost their lives in a natural disaster in the United States. Fortunately, her wrath evidently appeased, Marla weakened and passed over a sparsely

populated area of Mexico before being reduced to a minor disturbance as she forced her way across to the Pacific.

◆

NOVEMBER 21, 2011.
MIDDLEBURG, VIRGINIA.

Sam flipped off the television after she and Jason finished breakfast. They were becoming numb to the devastation that had
dominated television since Marla's landfall. The storm had left
Florida and the nation reeling. Given the number of smaller
storms that had pounded Florida and the Eastern Seaboard over
the past decade, private insurance companies had withdrawn
coverage from the areas at risk. The federal government had to
step in to insure those that otherwise could not obtain insurance.

But the strain created on the economy by the oil shock had
already dipped the country into a serious recession. While the
destruction might, ultimately, pull the southeast out of recession
due to a boom in construction to rebuild the hundreds of billions of dollars worth of property destroyed, the strain on the
Treasury would be enormous. The attacks on the solar power
satellite program being waged by Congressman Gallager could
only intensify. Sam could see the lines of worry that had begun
to crease Jason's forehead. She also felt him fidgeting in bed on
many nights and heard him get up to work on his computer in
the middle of the night.

Things were not going well for the Project. The President's
support slipped rapidly despite the successes of the first solar
power satellite and despite the clear progress being made on the
moon and on the George Wright. Marla might be the straw that
broke the camel's back. To make matters worse, given the mag-
nitude of the storm, the topic of global warming was again
gaining popularity. While nobody thought that things were as
bad as Sam, Jason and a few others knew they were, the public
discussion of global warming did not bode well for the Project.

Sam hoped that no other signs of global warming would manifest themselves in areas where they might raise the public awareness of the problem.

Jason kissed Sam and little Jason and left for the office. Sam turned little Jason over to his governess and went back upstairs to ready herself for work. When she finished dressing, she returned downstairs, kissed little Jason goodbye and walked out the door. The trip to the office took only 15 minutes.

She had set up an appointment with her father to discuss a bold, but risky maneuver. She and a few of her key assistants had run the numbers over and over again. They appeared to work. She stared at the numbers printed out in front of her one more time. She then folded them up and carried them over to her father's office.

Mr. Whitlock sat behind the same leather inlaid desk that Sam remembered hiding under as a child. She smiled at the thought of those days as she approached the desk. Her father glanced up and then looked at his watch. He stood up and began to walk around his desk to greet her. She met him halfway. They exchanged a brief hug and a kiss.

"Good morning, Samantha. How are you today?"

"Great, Daddy, just great."

"And how is little Jason doing?"

"Great! He's into everything. He's finally over the cold he picked up in preschool last week. And he's very excited about going over to Grandpa's house for Thanksgiving."

"I'm looking forward to it. But," rubbing his stomach, "I may have to cut back this year. I think that I'm gaining a little weight again."

"Don't be ridiculous, Daddy. You're in better shape now than you were ten years ago."

"I know. I know. But I want to be around to teach little Jason all of the important things in life like fishing, skipping stones and throwing a football." Sam frowned. She almost teared up remembering those early days with her father and the fact that, despite all of her and Jason's efforts, her father and little Jason would probably never have the opportunity to go fishing in the ocean like she had years before. "What are you frowning about dear? I know things are changing, but humor an old man. Besides, I have the right to dream, don't I? Now, what is it that you've been so busily working on lately?"

"You know how the political winds are blowing. Things are looking bad for Project Runaway. And because of his support of the solar power satellite program, it looks fairly clear that the President is going to lose his party's nomination next year to Congressman Gallager."

"I know. It's one of the reasons that my foray into politics was so brief. I got frustrated and refused to compromise my beliefs. As a matter of fact," Sam recognized the look in her father's eyes and thought to herself, oh no, not the speech about his tenure in Congress again, but her father focused his gaze back upon her, "Well, you know that story. I've told it to you a hundred times. Go on. Don't let me sidetrack you." Sam smiled.

"Well, I've spent the last several weeks looking over our financials. I think that we can step in and buy out the government's share of the solar power satellite operation. We don't need public funds to do it. We can make it work on our own."

Sam's father looked seriously into her eyes. "Are you sure about this, Sam?"

"Look," Sam spread several sheets of computer printouts across her father's desk. "From the projections we've run, the existing satellite is already returning a great cash flow. We already own all of the land that we need. In addition, given all of the insights that Jason has fed us over the past few years, we've decimated our competition. If we sell these subsidiaries, we should be able to raise enough capital to pull it off."

"But those have been some of our most profitable arms over the years."

"I know. But given what we know is going to happen over the next few years, they will lose profitability rapidly. We should sell them now, regardless of my plan. Besides, it will keep Justice off our backs. We've been too successful. I've heard rumors of a

Justice Department antitrust investigation. The divestitures should end that."

"But the announcement will cause our stock to plummet. We won't be able to borrow a penny. What you are talking about doing is something that the private sector has never undertaken before. Besides, with oil prices falling, we will never make enough off of the sale of electricity to cover our fixed costs."

"But you forget. We have the inside track. Given the current public perception and the climate in Congress, I'm sure we can execute a long-term lease with the federal government for the lunar base for a hundredth of what it would cost to create the base. The public and Congress have no clue how much work and money was secretly poured into the lunar base in the '80s and '90s and over the past few years."

"We'll get an unbelievable bargain. And, to top it off, we will still be receiving the covert support from Project Runaway. In addition, if the base becomes a private operation, we will not have the public scrutinizing the base. One of the thorniest issues Jason and the others face is how to explain how the base had come so far so quickly. They have been doing a great job emphasizing that the work has been done, in part by our company, under budget and ahead of schedule. But far too much has been accomplished in too short a period of time. As soon as the first astronauts return to the moon, we'd have a real problem."

"However, if we contract with NASA to send up our own people and keep the public out. Sure, the coverage will be enormous, but we can sell the right to televise the landings to the networks and can limit what is filmed. Once the mass driver is up and running, which should be within six to ten months, we can accelerate construction of solar power satellites and sell the power to local utility companies as well as other nations, if necessary."

"But what about the other nations? There has already been a public outcry over the nationalization of lunar resources and the violation of the Outer Space Treaty and the Moon Treaty. How are we going to deal with the international pressures?"

"How are they going to stop us? The United Nations has no

teeth. It can't enforce the treaty against the United States. Besides, the United States never signed the Moon Treaty. What are they going to do: embargo the United States? They couldn't do it. It would never work. All they could do would be to lobby public opinion and Congress to cut back on any grants that we might have or to cut out our access to NASA transportation. But until President Hammond is voted out, they won't be able to override his veto. We can do it."

"Have you forgotten that we are a public company? We will need to make disclosures to our shareholders."

"That's one of the beauties of the entire scenario. Once we announce that we want to buy the government's stake in the solar power satellite and lease the lunar base, we'll be labeled as idiots. Our stock prices will plummet."

"Yes. And our shareholders will clamor for us to be fired. Even though I control forty-five percent of the outstanding stock in the company, I doubt that I'll maintain control for long."

"I know, but between you, John Stabler and Charlotte Lillard, you control sixty-one percent of the stock of the company. If we can convince them of the merit of our plan, we can buy up the outstanding stock as the shares plummet and take the company private! It will work. It has got to!"

Stephen Whitlock looked at his daughter and then looked at the figures that she had been pointing at. He studied them for several minutes in silence. Finally, he looked back at his daughter and grinned. "You know, it might just work. But by God, what a risk we'll be taking! I took a lot of chances building this company, but none of the scope or scale that you are talking about. I never knew what you were really made of until now."

Samantha blushed. Her father had never been one to heap praise upon her. "Really, Father, what choices do we have? You would have come up with the same thing if you had focused on it."

Her father shook his head. "I can't say that I would have. Leave me alone with these figures for a while. I'll study them and get back with you as soon as I can. But, dear, either way, I am very proud of you."

"Thank you, Daddy." Sam leaned over her father's desk, gently kissed the top of his head and then turned and walked out of the office.

✦

THE FARM.
BLUE RIDGE MOUNTAINS, VIRGINIA.

When Jason arrived at the Farm's first checkpoint, the usual Marine guard was not on station. He assumed that the guard was sick. But as he drove closer to the main building, he noticed more guards than usual patrolling the grounds. They were not Marines.

He parked his car and entered the building at a brisk walk. Even the guard stationed at the front desk was different. Rather than a Marine sergeant's uniform, the man wore a black and gold uniform with a black beret. He came smartly to attention as Jason walked past him and did not even question Jason for his identification. Evidently, even though he was new, the guard had been thoroughly briefed.

Jason bypassed his office and walked into Lt. Colonel Bob Ryan's office. A captain in the same uniform as that worn by the guard at the front desk stood in front of Bob Ryan's desk. They both looked up as Jason entered the office, unannounced. Jason looked over the captain and then looked back at Bob.

"Bob, what the hell's going on around here? There are new guards all over the grounds in uniforms that I don't recognize. Who are they?" And, turning to the captain, "Who the hell are you?"

"Settle down, Jason. Didn't you get the memo I sent you last week? Apparently not. This is Captain Anthony Healy."

Captain Healy came to attention and extended his hand towards Jason. "Tony, Mr. Graham. It's a pleasure to meet you." Jason shook his hand.

"Now, Captain Healy, if you'll excuse us." The Captain saluted and exited the office. Jason took a seat across from Bob Ryan.

"Jason, it's not like you to overreact."

"I know, Bob. I've been a bit frazzled lately. What's going on? I am the head of this Project. I should have been informed personally of this change."

"You know me, Jason. I would have. But I was called away last week to meet with the General out at Vandenberg. Evidently, my assistant didn't get the message to you."

"Evidently not."

"Jason, you know as well as I do that we are in for some rather serious conflicts over the next decade. We decided that we needed to develop a special elite group within the military to support us when the need arose."

"General Maxwell initiated the training and selection process before President Hammond entered office. Since then he managed to convince the President of the need for such an elite military force. The President approved it. The new guards, led by Captain Healy, are a unit of the special forces. They are a multi-service group comprised of the best of the armed services. They call themselves the Black Berets."

"I can see the concern in your eyes. Don't worry. You'll see that we were right. One of these days we'll need these troops. They are the best trained and the best equipped in the world."

Jason shook his head. "But who do they answer to? I am the head of this Project, but I didn't know about them."

"Jason, the Black Berets aren't part of Project Runaway. They are funded out of a different black segment in the Pentagon's budget. Even the Joint Chiefs know very little about them other than their existence. None of the members of the Black Berets, other than their commanding officer based out of Vandenberg and Captain Healy, know anything about Project Runaway. As with the former Marine guards, they only know that this facility is connected with a highly classified project created under the direct authority of the President."

"But why were they brought here?"

"Straight? Rumor has it that Hamilton Smith and his crowd managed to pin one of the Marine guards down in a bar and loosen his tongue into describing the Farm. The guard even revealed that the President had been on the site on a number of occasions. The word is that Smith excoriated General Maxwell. The General, in turn, chewed out the Marine commander, had the Marines transferred off the base, and brought in the Black Berets. I understand that the poor Marine that leaked the information is now serving guard duty at a weather station in Antarctica."

"Why didn't you catch any of the flak? I thought you were the senior military man on this facility."

"Just lucky, I guess. Actually, while I am the highest ranking member of the military stationed here, I am a liaison officer. The Marine security detachment was under the direct control of the Marine commander, who in turn, took orders directly from General Maxwell." Bob leaned back in his chair.

Jason smiled. "You came out smelling sweet and clean as usual. It reminds me of the time that you ended up with Jenny in your room and Cynthia in mine. I still don't know how you got out of that one with your skin intact."

Bob smiled broadly. "Those sure were the days; weren't they?" Jason smiled back. "They sure were. Anyway," Jason slapped his hands against his legs and stood up, "next time, please try to give me some notice. Now, I've got to get back to work."

CHAPTER 21

MARCH 17, 2012.
THE WEDDELL SEA, ANTARCTICA.

Japanese fishing trawlers are reporting a 30% reduction in the harvest of krill this season. In addition, a research vessel studying penguin rookeries has reported a major decline in the number of hatchlings of all types of penguins, but Emperor Penguins, in particular. If the trend is not reversed, the Emperor Penguin will soon be added to the list of endangered species.

Scientists speculate that the decrease in both the krill catch and the Emperor Penguin is due to a combination of the increase in ultraviolet radiation reaching the surface through the ozone hole and the decrease in the salinity of the water due to ice melt.

DEER VALLEY, UTAH.

The skiing had been awesome that day. Despite the lack of snow early in the season, several storms had deposited over two feet of new snow the week before Jason and Sam arrived. The plane had barely been able to land at Salt Lake City due to the storm.

Jason and Sam were spending the week in Sam's father's house in Deer Valley to celebrate their anniversary. As much as they missed little Jason, they were glad to have the time alone.

Jason watched Sam slip into the hot tub as he finished mixing their drinks. He walked out onto the deck, set the drinks down, slipped off his robe and slid into the water next to Sam. He handed Sam her drink and then took a sip from his. The water felt great to his sore muscles. He used to be able to ski all day and party all night, but he was worn out. Between work and little Jason, he was no longer in the shape he'd been in the last time he had gone skiing.

A gentle breeze blew over the deck, pushing the steam off to one side. Jason looked up at the clear night sky. This high up and this far away from any city lights, Jason could just make out the Milky Way. After a few minutes of looking up, he noticed two new stars spaced close together. They looked out of place. He pointed them out to Sam. She smiled. "Honey, those are the first two solar power satellites. If you look closely, about one-half a degree to the left of the second one, you will see a dimmer spot where the third satellite is under construction."

"I see it. You know, I hadn't realized how long it's been since I'd looked at the night sky."

"You couldn't have seen them from Washington anyway. The satellites are in geosynchronous orbit over New Mexico and Arizona. But come next year, we should have something to see. Now that the first mass driver is operational, construction will really begin to accelerate.

"We've already purchased remote tracts of farmland in central Pennsylvania and Georgia for the construction of rectennas. They'll receive power from the East Coast satellites that we are about to build."

"You know, that idea of yours to lease the lunar base and buy out the government was brilliant."

"Thanks, dear." Sam leaned over and kissed Jason.

"I'll never forget the look on Congressman Gallager's face when the President announced the plans. It totally deflated the

Congressman after he had harangued on the House floor for hours, lobbying against future expenditures for "Hammond's folly." The money was needed here, at home, to care for the refugees of Marla and to prop up our failing economy. Besides, given that oil prices had fallen below the level that had existed before the Iran-Saudi war, the justification for the late President Wright's noble idea had disappeared."

"His jaw practically hit the floor when the President announced that he had cut back spending on the program to nothing and was actually raising revenues for the Treasury by selling the satellites and leasing the lunar base to Whitlock Industries. The move caught Gallager totally off guard and raised President Hammond's approval rating by six percentage points."

"I know. I remember. Too bad the boost in popularity faded so quickly. I'm afraid Gallager's going to win the Democratic nomination this summer."

"At least your old boss looks like he has a good chance of winning both the Republican nomination and the election in November. If he doesn't, the Project will go back into deep cover with lower levels of funding and..."

Sam slipped her hand onto Jason's leg and ran it upwards. "Jason, don't forget. We are on vacation. Let's quit talking shop."

"But..."

Sam put a finger to her lips, gave it a slight kiss, and then gently pushed it against Jason's lips. He kissed her finger and watched as Sam slid her head slowly under the bubbling waters of the hot tub. He smiled as he realized he wasn't quite as tired as he'd thought he was.

❖

VANDENBERG AFB.
CALIFORNIA.

"You've done one hell of a job, General." Congressman Friedman looked at the dots of light representing solar power satellites on the main board of the Vandenberg AFB launch control room.

"Thanks Joel, but this is just the beginning." General Maxwell led Congressman Joel Friedman out of the control room and down a corridor to an elevator. The elevator descended for roughly twenty seconds before coming to a halt. The doors opened to a small waiting area. The security guard, a Black Beret corporal, saluted as the General and Congressman Friedman entered the room.

"Corporal, is the train here?"

"Yes, sir!" The corporal stroked a few keys on the computer terminal in front of him. The Congressman noticed a slight breeze as a set of doors on one side of the room opened.

The General extended his hand: "After you, Joel." Congressman Friedman entered the room on the train that consisted of roughly a dozen seats. He took a seat at the front of the train. The General sat down next to him. Congressman Friedman knew what to expect from his prior rides in a maglev train at the Chaco Mesa power station. The train rapidly accelerated away from the station. The Congressman knew that it would begin to decelerate in less than five minutes and would arrive at its destination in about 10 minutes.

"Are we secure in here?"

"Of course Joel. These things have been tested..."

"That's not what I meant. Can we talk?"

The General's face reddened slightly. "Yes, it's secure."

"Good. Now, have you taken care of the construction crew?"

"Not yet, they're finishing up this week. Are you sure about this Joel? You know we're talking about 30-40 lives."

Congressman Friedman scowled. He'd been over this with General Maxwell several times already. As Jason Graham had pointed out, when things got rough, elements of the military might turn in order to obtain a safe place to live. They had to keep the undersea base a secret for all but those select few that would have a place in the facility or in the similar facility being built in the Gulf of Mexico. Joel allowed himself a smile because he alone knew of the existence of both sites. "You're not getting soft on me, are you?"

"No sir, Congr... Joel. It's just that it seems such a waste. They're all smart people. We could have a place for them."

"That's true. But, there are plenty of qualified people for the few slots available. We've picked most of them already. Besides, they're going to die anyway. We're just upping the schedule a little."

"But what if the President or the other Project Runaway team get wind of what we're doing?"

"They won't. Your people are the best and they are trustworthy."

"Yes they are. They are the best of the best. The Praetorian Guard, hand picked out of the Black Berets. They'll get the job done."

The General still looked tentative. "But Joel, even the best plans can be discovered. What then."

Joel looked into the General's eyes. He saw doubt. The General was getting old. A few years ago there would have been no doubt. "If any low visibility person finds out, we'll plug the leak. If we can't, what can they do? They can't bring us to trial. We'll threaten to go public. If it happens later in the game, when they don't care about publicity, it will be too late. We'll have the power to stop them."

The train arrived at the undersea base off the California Coast. Another Black Beret guard greeted the General and Congressman Friedman.

CHAPTER 22

AUGUST 10, 2012.
CHICAGO, ILLINOIS.

The Chicago Board of Exchange temporarily halted the trading of corn futures following the Secretary of Agriculture's announcement of the disappointing fall harvest. The heat wave and drought in the corn belt caused far greater damage to the corn crop than previously believed. The result will be a significant increase in food prices for consumers this winter. Not only will corn and vegetable oil prices increase, but feed grain prices will also rise and may result in an initial decrease in the cost of meat as farmers trim their herds to reduce the cost of feed, creating a glut on the meat markets. However, the glut will only be temporary because it will take the ranchers years to replenish their stock. Prices of meat will rise dramatically next year.

THE FARM.
BLUE RIDGE MOUNTAINS, VIRGINIA.

Jason arrived in his office, hung up his jacket and told his computer to turn itself on. He scanned the headlines on the screen. In addition to automatically reporting any data that might be relevant to global warming, he had programmed the computer to scan the Web and summarize stories or reports on other topics of interest to Jason. Problems were beginning to surface that could lead to a public revelation of the runaway greenhouse effect. It was about time for them to start the second phase of their disinformation campaign.

As head of the Project, Jason did not have to keep up with the distasteful task. Jason had plenty to do reviewing and revising the programming for the space city's operations. He knew that Jeff Nakamura, Arthur Jacobson and the others at EnviroCon could handle the project.

As Jason poured over lines of code on the screen, a bell chimed. Annoyed. "Yes."

"Telephone call, Mr. Graham."

"I instructed you not to accept any calls." Jason was glad that he'd begun to use the voice-activated feature of his computer. It let him vent his anger at the machine.

"You did, but you also programmed me to interrupt you for certain callers if they indicated that the matter was urgent."

That caught Jason's attention. "Who is it?"

"Your wife."

"Put her through."

Sam's face appeared on the screen. From the view behind her, Jason could tell that she was at home. "It's not something with Little Jason?"

"No, dear, no. It's about work."

"Why are you still at home?"

"They caught me just as I was leaving. I decided to call you from here rather than the car." Hamilton Smith had installed secured lines between the Farm and Jason's house and Sam's office, as well as between several other locations, but was unable to do so with car phones. The lines were monitored. If they detected any unauthorized monitoring of the lines, the call would automatically be terminated. That had happened a couple of times, but Jason never knew why.

"What is it?"

"The Catcher has gone awry."

Jason ignored Sam's attempt at a pun. "What do you mean, 'gone awry'? It's been working perfectly for several months now. I designed most of the software myself."

"I know. That's why I'm calling you. Jamie Hampton called me from the George Wright to tell me about the problem. They've had to shut the mass driver down. The Catcher failed to confirm that it was receiving the payloads. Several tons of raw materials are possibly heading off into oblivion right now."

"Let me call Hampton up. I'll talk with him and get right to work on the problem."

Jason noticed a smile crease the corner of Sam's mouth, although he wasn't sure why. He ignored it. "No, Jason. Hampton insisted that you get up there to work directly with their computer. I've already booked you on a flight to New Mexico."

"But why?"

"I don't know. But I trust Hampton. I'll meet you at Dulles with a small bag. See you soon."

The screen went blank when Sam disconnected the line. The computer chimed back in thirty seconds after the line disconnected. "Do you wish to return to the point where your work was interrupted?" Jason hadn't realized that thirty seconds had elapsed. He had been thinking about the problems that could have arisen on the Space Station. "No. Please download my current models on the 'Catcher' onto disk and then shut down. Also, if anybody calls, I'll be out of the office for several days."

A few seconds later several CDs popped out of the drive on Jason's computer. He put them in his briefcase with his laptop and headed for the door. It would take him longer than Sam to get to Dulles. He nodded to several Black Berets that he passed on his way to his car.

Thirty minutes later he arrived at Dulles International Airport, spotted Sam's green Land Rover and pulled in next to it. Sam had given up the Jaguars now that they had a child and were living in the country. Jason stepped out of his car and walked over to Sam's. Sam got out and gave Jason a hug and a kiss. Jason was surprised to see Kim Huang walk around the back of the car.

Sam, noticing Jason's surprise, said: "I brought Kim with me to drive your car back to the house." As was frequently the case, Sam was one step ahead of Jason. In his rush, he'd forgotten that senior project members were not supposed to leave their cars in public parking lots. If Sam hadn't remembered, Jason would have had to deliver the car to Hamilton Smith's team upon his return for a thorough examination.

Kim handed Jason his small, black leather overnight bag. Sam then handed Jason a ticket and began talking rapidly. "Here's your ticket. I booked it through the company. You'll only have a short layover in Santa Fe. I've arranged for a company helicopter to fly you out to the Chaco Mesa spaceport."

Jason took both of Sam's hands in his own. They were shaking. "Sam, don't be so nervous."

"Nervous? I'm not nervous."

Jason smiled. "I know you better than that. I'll be fine. We've both known all along that, eventually, we would be traveling into space. Little Jason will as well."

"I know. But this is the first time. I can't help but worry."

"It's not as if I'm going up on one of the old Shuttles or the Delta Clipper. The ship I'm going up on is a glorified jet. They've been flying for years on the edge of space in flights across the Pacific Basin. You know as well as I do that only minor modifications were required to convert the jet into a spacecraft."

"I know. But I wouldn't be your wife if I didn't worry a little." Jason took Sam in his arms and gave her a firm hug and then a kiss. Before pulling apart, Sam whispered into Jason's ear, "I love you."

The flight out to Chaco Mesa was uneventful. Jason spent the time reviewing his models on his laptop. While the laptop had sufficient speed and memory to let him review the programs, it would have taken far too long to run the programs on the computer.

When the helicopter arrived at the Chaco Mesa complex, the changes surprised Jason. In addition to the rectenna field, a large compound had developed near the site of the old public relations building. Furthermore, a couple of miles away from the public compound, Jason saw several very long runways as well as several large tanks containing what Jason surmised were liquid oxygen and liquid hydrogen to fuel the hypersonic scramjets.

Several years before, Whitlock Industries had secured a contract from NASA to modify the hypersonic jets now being used by several airlines for transatlantic and transpacific flights as a replacement for the old Space Shuttles that were nearing the end of their useful lives. Thanks to additional funding provided by Project Runaway, Whitlock Industries had managed to deliver the new aircraft/spacecraft on time and under budget. The modified aircraft could now ascend all the way to low earth orbit and the Space Station Freedom.

The helicopter landed Jason at the new spaceport. He entered the waiting area and checked in at the main desk. Whitlock Industries had built the facility under a contract from NASA and then leased it back from the government. It currently served as the principal spaceport for passenger travel to and from the George Wright and the Whitlock Industries orbital construction sites; although it also served as an emergency backup for NASA. Whitlock Industries had also built a facility for NASA near Huntsville and for the Air Force in Colorado. Jason had lobbied hard for and strongly approved NASA's decision to locate its new spaceport in Huntsville to avoid the vagaries of Florida weather that, with increasing frequency, had delayed launches. After Marla, what had been declared to be a waste of money by the Gallager clique in Congress had become NASA's salvation.

Unfortunately, because of the nature of the Delta Clipper class heavy lift vehicle, they still had to be launched from the coast in the event of a catastrophic launch failure. The public still perceived the risk as being too great to launch them over populated areas. Fortunately, over the last few months, NASA's launch schedule of the heavy lift vehicles had diminished as a result of its largest customer, Whitlock Industries and the solar power satellites, shifting to using raw materials from the moon. Now, Jason was concerned that supplies of raw materials to the orbital construction crews might again become reliant on NASA and the fleet of heavy lift vehicles if he couldn't fix the problem with the Catcher. That could be a serious problem given the increasing number of violent storms at the Cape and in Washington.

Jason looked out of the window of the terminal at the sleek lines of the spacecraft. He could see small clouds and ice forming around the fuel intake valves from the super cold fuel being pumped into the spacecraft's tanks. When the boarding announcement came Jason got in line with a number of well-dressed men and women that were heading up for their construction shifts. Whitlock Industries psychologists had developed the construction schedule to minimize psychological trauma and maximize productivity. Four separate construction crews were based at the station at one time in order to keep work moving along in four daily shifts. Each crew would have 6 hours on and 18 hours off. The construction crews would work for forty-five days and then be relieved for a sixty-day vacation on Earth.

Seating in the spacecraft was similar to that of any airliner: an isle down the middle and two seats on either side of the aisle. Jason received an aisle seat fairly close to the front of the craft. A man in a blue business suit sat down next to Jason in what would typically be the window seat. Unfortunately, due to the design of the craft, there were no windows for the passengers. Jason introduced himself to the man seated next to him. The man, David Mahoney, explained that a hotel chain had hired him to negotiate a contract to manage a small hotel on the George Wright. Before they got beyond their exchange of initial pleasantries, a voice over the intercom interrupted them.

"Ladies and gentlemen, welcome aboard the Von Braun. FAA regulations require that you watch the video displays located in the back of the seat in front of you. The video will run until we are prepared for takeoff. If you have any questions please press the button on the armrest of your seat. You will receive a response over your video display."

"As we take off, the displays will switch over to the cameras mounted in the flight deck and on other areas of the spacecraft. You will be able to view the takeoff and landing on your video displays. No food or beverage services are provided; however, a zero gravity toilet is located in the rear of the spacecraft. We recommend that you wait until we reach the Space Station Freedom complex to use the rest room if possible. The video will cover the remainder of the matters you will need to know, including what to do in the event of motion sickness. Have a pleasant flight."

The video kicked on as the intercom cut off. Jason began to watch the film. Several minutes before takeoff the video switched over to an external view of the Von Braun. Jason felt the craft lurch and then watched as the craft was towed to launch position. The takeoff was very similar to that of a conventional aircraft until the craft reached 40,000 feet. At that point Jason felt a surge of power as the scramjet engines cut in.

Twenty minutes later, the engines shut down. They coasted into orbit and into weightlessness. Jason felt a peculiar sensation in his stomach, but he did not become ill. From what he could hear, not all of the passengers were so lucky. Once in orbit the orbital maneuvering rockets took over to gently nudge the craft into its rendezvous with the Space Station Freedom complex.

From the video display, Jason watched as the Von Braun approached the orbiting complex. At first, all Jason could see was a speck of light set off from the curve of the Earth that dominated the picture. Slowly, the speck grew into the Space Station Freedom Orbital Complex. The complex looked less like a space station than a sky scraper under construction. Girders and beams extended out from the pressurized modules at the center of the original space station. Off to one side orbited an antenna farm and a complex of numerous old shuttle external tanks.

The screen split on final approach to show the view from the ship and the Station simultaneously. The ship finally docked at a port next to the cluster of shuttle external tanks. The air lock in the front of the passenger compartment opened and several people in blue overalls entered the craft. Jason began to unbuckle his seat belt until he remembered to put on the velcro slippers that the video had mentioned and that he found in the pouch in the rear of the seat in front of him. He put on his slippers, planted his feet firmly on the floor and released his seat belt. Then, holding onto the seat, he walked out into the aisle and off of the spacecraft.

A surprising number of people had forgotten the velcro slippers and were floating about the spacecraft. The men in overalls went to their assistance with mild disdain. Jason was glad that he didn't require their assistance. Once in the Space Station, he found himself in a large room partitioned out of an old shuttle external tank. Evidently numerous of the passengers had been on prior construction crews and knew the routine. Jason followed them. They went to an open air lock on one side of the room. It led to a passageway that led to another open air lock. Jason noted several closed air locks along the way, but none of the other passengers had stopped at any of them.

After passing through the last air lock, they entered another large room filled with chairs. Most of the other passengers took their seats, but a few removed their velcro slippers and began leaping through the air. Several individuals in the room had on the same blue overalls as the people that had first entered the Von Braun from the Space Station. One of them was standing behind a desk working on a computer terminal. Jason walked over to her.

She had closely cut hair and a puffy face. Jason could see from several of the women passengers why she kept her hair short; one blond with shoulder length hair was trying to pull her hair down, but it kept floating up over her head. The sight was rather comical. The puffiness was probably due to the redistribution of body fluids as a result of weightlessness. "Excuse me, I just arrived on the Von Braun. How long will it be until we transfer over to the ship that is going to take us up to the George Wright?"

The woman looked up, perturbed. "Can't you see that I'm busy." She returned to her work.

"I hate to bother you, but I'd..."

"Look, Mr. . .."

"Graham."

"Mr. Graham. I've got a job to do here preparing for the rendezvous of the orbital transfer tug. I don't know how you got up here without passing your training course, but I don't have time to answer your questions." She went back to her work.

Jason started to interrupt again when he noticed David Mahoney had entered the room. David saw him and signaled to him. "Jason, glad to see a familiar face again."

"Me, too, David."

"I met a few of the other passengers during the orientation classes. But, because I'm not one of the construction crew or the permanent base personnel, they seem to look down on me. How about you? I didn't see you at any of the orientation classes and assumed you'd been up before."

"No. It's my first trip and I didn't go to an orientation course."

"Didn't go? I thought you couldn't get up here without one."

"I came at the last-minute. I'm a consultant for Whitlock Industries and needed to come up unexpectedly." Jason obviously couldn't go into any further detail about his mission. "Say, you may be able to help me. I tried to get some answers from that lady over there," Jason motioned to the woman behind the desk, "but she gave me the cold shoulder. What do we do next? When do we transfer to the OTV?"

"We don't. This is it."

"But..." Jason stopped and began to think. David continued.

"I'm surprised you don't know, but I'll tell you what I learned in the training class. When we got to the Station, we walked over to another external tank that has been modified to take us all the way to the George Wright. This—," David extended his arm to point to the room they were in, "—is the passenger section of the transfer tank. The rest of the old external tank has been modified to carry supplies and fuel. The OTV or Tug, as we called it in class, will drop off a similar external tank on the other side of this Station with the returning construction crew and any other passengers coming back from geosync or from the Moon."

"Those passengers will transfer to the Von Braun for the trip back down to Chaco Mesa. The Tug will then disengage from that transfer tank and fly around the Station to hook up with this transfer tank. Once it's hooked up," David looked down at his watch, "and that should be within an hour, we will be uncoupled from the Station so the Tug can boost us up to the George Wright. It's fairly simple, actually. And from what I hear, very economical."

Jason thought out loud. "I should have thought of it. Why build a special vehicle to transfer passengers and one for cargo when you don't have to? It makes perfect sense."

"You catch on fast."

Jason shrugged. "What about seeing the rest of the Station or going over to the original sections of the Station?"

"I'm afraid we're not allowed to do that. It would create too much disruption in the experiments conducted in the labs around the Station and would make it far too difficult to keep track of all of the passengers. Unfortunately, we're required to stay here once we arrive. They did, however, rig up a virtuality tour of the entire Freedom complex if you want to check it out." David pointed out to one corner of the room that Jason had not noticed before. Several people were seated in chairs with the familiar helmets and gloves used for virtual reality work. "It was designed originally for training but has been converted to recreational use."

"I may try it out, but what about you?"

"I think that I'll try my hand at flying before we leave the Station."

"Flying?"

"Yes. It's one of the recreational activities that the Portman hotels will be marketing if we get the rights to put a hotel on the George Wright. Just look at all of the people floating around, having a great time. Our marketing people think it will be a big hit." David pulled out a set of gloves and flippers from his bag. "The company designed these. They want me to try them out up here. If I don't try them now, I'll have to wait until I can schedule a time for one of the zero g rooms on the George Wright. We'll be under moderate acceleration most of the way to the Station and I doubt that these will work in more than a twentieth of a g. Want to try? I've got an extra set."

"Sure." He handed a set to Jason and Jason put them on. David launched himself into the air a bit too briskly and bumped his head on the ceiling as he tried to use the flippers to swim away. Jason, learning from the example, pushed his arms down gently at his sides and floated up into the air at a slower pace. After a little practice, both he and David had the hang of it and were looping and spinning through the air. The others in the room couldn't change direction once they pushed off a surface. But, due to the gloves and flippers, Jason and David could. Soon, most of the eyes in the room had turned to them.

Before he knew it, the intercom announced that they would be leaving the Station in ten minutes. Jason and David made their way back down toward where they had left their bags. Several of the other passengers came over to talk with Jason about his "flight." He referred them to David. A crowd had gathered around David and he had a smile on his face as he described flying and the concept for the recreational hotel.

Jason walked over to try out the virtuality tour but found out that they had been shut down when the launch announcement was made. He returned to his chair and prepared for the launch. Several minutes later, Jason felt a slight tug as the transfer tank was pulled away from the Station by the Tug. Shortly thereafter he could feel a slight pull towards the floor as the Tug began to accelerate. Several loose objects drifted slowly back down to the floor of the room as the acceleration continued.

The trip out to the George Wright in geosynchronous orbit would take eleven hours. Jason decided to try to take a nap. He figured out how to recline his chair, did so, strapped on the seat belt in the event that acceleration stopped, and shut his eyes. He awoke several hours later feeling very refreshed. Evidently what he had read about sleeping in zero or low gravity was true; you did get more benefit from less sleep. Although he was beginning to feel congested, he felt fortunate that he wasn't suffering from motion sickness as were close to half of the others on the flight.

Jason got up from his chair and, with his first step, pushed off the floor and practically hit his head on the ceiling. He had forgotten that they were under very low gravity. When he finally floated back to the floor he gingerly walked back to his chair and put his velcro slippers back on. He then made his way to the bathroom. Fortunately, due to the slight acceleration of the vessel, he didn't have to use the "oh gee whiz" attachment, as the popular press called the suction device designed by engineers to enable men and women to go to the bathroom in zero gravity.

When he returned to the large room, he went over to a counter where he saw several people getting a drink and a bite to eat. He helped himself to some water and an apple. Then he went back to his chair to watch the screen that had been lowered against one wall while he'd been asleep. It showed the approach of the Tug toward the George Wright. While they were still several thousand meters away, an announcement came over the intercom requesting that all passengers take their seats due to the necessary docking maneuvers. The passengers complied. The mild acceleration of the craft quit, briefly returning it to zero gravity. Jason then felt several harder jolts as the chemical maneuvering engines took over from the ion drives to nudge the ship gently in toward the docking area of the station.

Unlike the Space Station Freedom complex, the George Wright looked like a Hollywood space station. It had the classic wheel, spokes and hub configuration. The wheel rotated to generate artificial gravity. A section of the hub also rotated, while another section, where the Tug would dock, remained stationary. Construction was about to commence on the second wheel of the Station in order to double the number of construction workers that could be housed on the station from two hundred and twenty to four hundred and forty.

Off to one side of the station, Jason could just make out one of the operational solar power satellites glistening in the sunshine like a jewel. One of the beauties of the station was its ability to move from one location in geosynchronous orbit to another using several large ion engines attached to one end of the hub. Because the solar power satellites supplied the power, fuel didn't enter into the calculations for a move. Once the inconvenience of transporting construction crews to the solar power satellites became too great, they would move the station to a new location where new satellites would be built.

After what seemed like an eternity for the Tug to move the external tank the last hundred meters toward the station, Jason finally felt the slight jolt as the tank hard docked with the station. About ten minutes later an air lock opened and several people in brightly colored overalls entered through the air lock. An announcement came over the intercom directing the construction crews to exit first. Jason and a few other passengers that, evidently, were not members of the construction crew waited while the bulk of the passengers filed out the air lock.

After the last construction crew member exited the air lock, the intercom requested that all other passengers enter the Station. Jason, David Mahoney and several other passengers made their way onto the station. As they did, they entered a large, brightly lit, white room. Data screens located along the back wall drew Jason's attention. Jason noticed his name on one and walked over to it. A young lady, Jason guessed she was in her mid thirties, about five foot six, with short-cropped chestnut brown hair, wearing bright green coveralls waited below the screen. Jason walked over to her.

"Mr. Graham?"

"Yes."

"I'm Jamie Hampton. It's a pleasure to meet you." The commander of the George Wright extended her hand towards Jason. After a moment's hesitation, Jason extended his own. Now Jason knew why Sam had smiled slightly at the name of Jamie Hampton. He had assumed that Jamie was a man and Sam, knowing his prejudice, had decided not to tell him that Jamie was a woman. He smiled to himself. One of these days, he'd get even with Sam.

"My pleasure, Ms. Hampton. And, please, call me Jason."

"Only if you call me Jamie."

"It's a deal."

"As much as I'd like to give you a tour of the station, we need to get right to work. I'll brief you when we get to my office. If you will, follow me."

Jamie began walking toward a door on one side of the room. "Oh, and by the way, please be careful. I sometimes forget how difficult it is to adjust to the change from zero g to the $\frac{3}{4}$ g gravity that is maintained in the Wheel. You may feel a bit of discomfort moving through to the elevators in the rotating section of the Hub because of the speed at which it's spinning, but any discomfort you feel should disappear as we ascend up the Spoke elevators towards the Wheel. We have found that ninety percent of the population can handle the spin necessary for $\frac{3}{4}$ gravity without discomfort and that $\frac{3}{4}$ gravity is enough to counter any serious loss of bone or muscle tissue."

As Jamie was speaking, Jason followed her out of the docking area of the Hub and into a relatively narrow corridor that seemed to wind its way upwards. Jason felt pressure begin to increase on his feet and legs as they walked. Jamie frequently glanced over at him and Jason smiled back. While he felt a little queasy, it didn't really bother him.

"I'm taking you to the passenger elevator in Spoke I. It will take us to the administrative sector of the Station. I hope you don't mind, but I wanted to brief you as soon as possible. After the briefing I'll have someone show you to your quarters if you need a rest; if not, I can have your bag dropped off there."

"I'm fine. I'm ready to start working right away, although I could use something to eat."

"I'm sure that we can manage that."

Finally the corridor opened up into a small room with two doors on one side and several benches to sit down on. Several other people, two dressed in blue coveralls and one dressed in red, were already in the room. A door opened.

"We got lucky. The elevator's here." Jamie signaled Jason to enter the elevator. It looked as if it could accommodate as many as ten people. The other people that had been waiting also entered the elevator. They exchanged a brief greeting with Jamie, but Jamie didn't introduce them to Jason.

Jason turned to Jamie and spoke in a softer than usual voice to keep from being overheard. "I expected that the elevator would have windows to look out at what must be a spectacular view."

"We thought about it but decided, for a variety of reasons, not to include any windows."

"One other thing I was wondering, why does everybody here wear different color coveralls?"

Jamie laughed. "They're very comfortable and practical. Why? Don't you like them?" she said with a smile.

"That's not what I meant."

"I know. You will be issued blue coveralls. Blue is worn by all engineering and scientific personnel. Grey stands for temporary visitors. You see very few grey coveralls. That's one reason that I have assigned you blue coveralls. You won't stand out as much. The green that I am wearing is the designation for administrative personnel. The construction teams will wear tan when they are on duty and dark purple when they are off duty. The black uniforms are for the security personnel."

"While we permit personnel to wear civilian clothing on the station, it is generally only worn in the recreational and residential areas of the station. In addition to making visible identification possible at a glance, the uniforms also have very small transmitters woven into the fabric. Each person's coveralls will enable them to gain access to the areas that they are permitted to be in and are tracked by the computer and monitored by our security team."

"That's all very interesting, but I'm surprised by your need for such tight security. After all, don't you run thorough security checks on everybody that comes up here?"

"We do, but you can never be too careful. Given the current political climate, both domestic and international, we don't take any chances. This facility is far too valuable, and vulnerable."

The elevator came to a halt and the doors opened. Jamie and Jason let the others on the elevator exit. Jason began to follow them off, but Jamie extended her arm to stop him.

"This isn't our level. We are on level two."

The doors shut and the elevator moved up and stopped. The doors opened and Jason followed Jamie off. After taking a few steps off of the elevator, Jason's eyes were drawn to the windows on the opposite side of a small lobby. The curve of the Earth slowly moved through the window. Almost absentmindedly, Jason walked across the lobby to look out the window. Stars now filled the window and continued their slow clockwise trek across the window. Jason could see thousands upon thousands of stars, far more than he had ever seen before.

Jamie walked up quietly behind him. "Breathtaking, isn't it?" Jason nodded. The moon had just passed by. "Even though I've been up here for months, I still stop and stare occasionally." They stood in silence as the Earth reappeared in the window in all of its blue brilliance.

Jamie tapped Jason on the shoulder. "I'll take you for an even better view to the observation deck later. But for now, let's get to my office. And, by the way, you can take off your velcro slippers now." Jason looked around for a chair and noticed several sofas positioned to maximize the view out of the window. He sat down in one, pulled his slippers off of his shoes and placed them in his overnight bag that he had been carrying. "By the way, I put mine in this pocket on my overalls. When you get your overalls on, you may want to put yours in the same pocket. The pocket was designed for the slippers. If you need to go back down to the zero g section of the Hub, you'll have your slippers handy."

They walked down a corridor along the outer wall of the station, periodically passing small windows until Jamie stopped at a door and opened it. They entered a large, brightly lit room filled with built-in desks containing computer terminals and work stations. About a dozen people, almost all in green coveralls, were working hard at the various work stations. Jamie walked to one end of the room and opened a door with her name on it. She motioned Jason to step into the room.

The room was not at all what Jason would have expected. In contrast to the stark white light and white corridors and modern furniture, he stepped onto a plush burgundy carpet and looked around at what appeared to be dark wood paneling. Jamie's desk was of a light, burled wood. A large navy blue leather chair sat behind the desk and a credenza matching the desk sat behind the chair. If not for the small window showing an alternating view of the Earth, the moon and the stars, Jason wouldn't have guessed that the office was in orbit.

Jamie walked around the desk and sat down. She motioned to one of two matching guest chairs. "Have a seat." Jason sat down. "I know it's not what you'd expect. But then, that's one of my privileges as station commander.

"Now, let's get to business. Catcher number one received its entire load on schedule and without error and is waiting to be unloaded. Catcher number two pulled into position as scheduled and the mass driver began shooting the raw materials off of the Moon on schedule. At first the Catcher reported that it was receiving the materials. But the figures for the mass received didn't correlate with the mass launched from the Moon."

"How much mass are we talking about?"

"Approximately a metric ton."

"A ton of mass! That's just not possible. I designed most of the software myself. The safety systems would have kicked in long before a ton of material went unaccounted for. It was set to shut the mass driver down if more than a pound went unaccounted for."

"We know."

"Well, where's the extra mass going?"

"We don't know. We haven't been able to locate it. But, as you know, the Catcher is positioned in such a manner that in the event the Catcher failed, the materials, which are launched off the surface of the moon at several thousand miles per hour, would end up in a harmless orbit around the Moon."

"I know, but if a ton of mass is missing, then the entire program must be flawed. It could be heading anywhere."

"We know that Jason. We also know that it is not within three degrees of where it was supposed to be. That's why we got you up here as fast as we could. Our programmers have not found anything wrong with the software. We thought you might be able to find out what went wrong."

"I can see why you wanted me to get to work right away. I don't like the idea of a ton of materials shooting around in an unknown orbit at thousands of miles per hour."

"You don't. How do you think I feel? If only a pound or two of that material were to hit this station, this station would be totally destroyed. The loss of life would be total. Whitlock Industries would never recover from the blow."

And neither would Project Runaway thought Jason. "But the odds of that happening are hundreds of millions to one. You know that."

"I know, but somebody could change them."

"What do you mean?"

"Sabotage; what else? It could happen. While I have no evidence of sabotage other than the failure of the Catcher, we can't afford to take the chance. Jason, I need you to find out what went wrong with the program and fix it. If it's sabotage, we need to find out as fast as possible. My programmers and the programmer stationed on the lunar base have been working on the problem, but one of them could be the saboteur. I needed somebody with your expertise that I could trust."

"Let's hope the problem is with my software or with the hardware and not a saboteur. Where can I work?"

Jamie got up from her desk and walked over to one of the wooden panels and pushed on it. "Right in here." Jason walked through the door into a much smaller office with a modest desk and chair. "This office belongs to my assistant. He's on leave right now. It has a terminal that will give you direct access to all of the files in the Omegatron and can be voice controlled or operated by the keyboard, whichever you are more comfortable with."

"I use both."

"Computer, this is Commander Jamie Hampton, the next voice you hear will be that of Jason Graham, he is to have full access to all files on my authority. Jason..."

"Computer, turn on the terminal and pull up the Catcher subroutine." The terminal came on, filled with lines of code. Jason sat down in the chair in front of the terminal. He pulled out the CDs that he had brought with him and inserted them into the CD drive. He then began to type on the keyboard.

"Jason, before you get too involved, do you need anything?" Jason stopped typing and turned in the chair.

"I'm glad you reminded me. Before I forget, could you send a message to my wife, Samantha Whitlock Graham?" Jason looked down at his watch, "She should still be at her office. Let her know that I've arrived and will call her as soon as I can."

It was Jamie's turn to smile. "Now I know why Samantha had so much confidence in you when I called her. I'll let her know. Anything else?"

"Yes. Could you get me a Diet Coke and a sandwich?"

"Sure." Jamie went back into her office as Jason hunkered down behind the terminal.

CHAPTER 23

Representatives of the 210,000 residents of the Maldive Islands have arrived in New York to request urgent support for the relocation of the Islands' population. The Islands narrowly avoided disaster when Typhoon Angie skirted them earlier this year. Given that the highest part of the Maldives is only six feet above sea level, the threat of serious storms cannot be understated. In fact, the islanders claim that large sections of the islands are already under water due to a rise in sea level over the past decade.

✦

Jason sat back in his chair and stretched his arms over his head. He happened to glance down at the table next to him and saw a sandwich, chips and a glass of soda sitting on a plate next to him. He hadn't noticed it being brought in. Jason picked up the sandwich and bit into it. Not knowing what to expect, he was pleasantly surprised by the ham and cheese with lettuce, tomato and mayonnaise. He wolfed down the sandwich, took a gulp of Diet Coke and stood up.

He walked over to the door into Jamie Hampton's office and knocked. The door opened in response and Jason walked into the station commander's office. A tall, dark haired man in black coveralls stood on the opposite side of the desk from Jamie with a scowl on his face.

"Jamie, I hate to interrupt, but I've got to." Jason paused and looked over the man in the black coveralls.

"It's all right, Jason. This is Juan Vasquez, the head of station security. Juan, this is the man I told you about, Jason Graham." Jason and Juan briefly shook hands.

"What have you found out, Jason?"

"I'm afraid it's not good. There's nothing wrong with the Catcher or the software. I also checked on the mass driver's software. It's working within design parameters."

"Are you implying that there's been a breach in security based on the fact that the software you designed is working as designed? How do you know your tests aren't flawed?"

"I entered the program through my back door. It enabled me to bypass the built-in security programs. I then ran the equivalent of thousands of hours of real time simulations without a failure. Besides that, I found the saboteur's calling card in the program. Oh, it was very well hidden. If I hadn't used my back door, the saboteur's program would have erased itself, but it's there."

"Do you know who did it?"

"No."

"Can you find out?"

"Just from what the person did in the program, probably not. But we've a far more serious problem than locating the saboteur."

"What could be more serious than that? He could cause far more damage if he's not discovered."

"I doubt that. He's already caused all of the damage that he needs to cause." Both Juan and Jamie stared at Jason in silence.

"If you know what he did, can't you fix the problem so that we can get back to work."

"I can, but I didn't. I left the program in place."

"Why?" Both Jamie and Juan asked at the same time.

"Because by leaving it there the saboteur will not know that we're on to him. We will be able to trap him into revealing himself."

"But how?"

"Because within the next twelve hours, the saboteur is going to be getting off of this Station." Jason paused to let what he had said sink in. "Several tons of raw lunar materials are at this moment in an orbit that will cause them to impact this station in," Jason looked down at his watch, "just over twelve hours and fifteen minutes." Both Juan Vasquez and Jamie Hampton looked stunned. Jason continued. "The saboteur temporarily changed the launch parameters for the mass driver. I've uncovered the true launch parameters and ran a simulation to project the trajectory of the raw lunar materials. The orbital calculations were quite complex and quite elegant. After looping through the Earth-Moon system several times, the matter will hit this station."

Jamie recovered from her initial shock first. "We'll never be able to evacuate the station by then. There are two hundred and seven people working here now." Jamie paused, her mind racing ahead. "We could transfer maybe twenty to the lunar base and cram another hundred on the external tank transfer vessel you came in on, that's assuming that Freedom could handle one hundred additional people, which is questionable. But that leaves eighty-seven people here. That's not acceptable. Could we move the station in time? Computer, calculate how far. . ."

"Jamie, don't bother, I've already thought of that. Unless the computer is wrong, we can't move the Station in time."

"But we could use the two Tugs we have in orbit to assist the station's own engines." Jason held up his hand.

"Jamie, I know the materials' orbit. When we finish, I'll call a few contacts in Washington. The materials will be destroyed.

In the meantime, didn't I see a Catcher off in the distance when I arrived?"

"You're right. Catcher I is over by the solar furnace dumping its raw materials."

"Move the Catcher into position in front of the station. I hope we won't need it as a backup, but we might. I'd rather be safe then sorry. Until we get a few larger stations built at the Lagrange sites, this Station is our basket; and all of our eggs are in it."

"No problem, Jason. I'll get the computer to work it out."

Juan began to smile. "I think I know how you're going to catch the saboteur, Jason. I like it!"

✛

AUGUST 12, 2012.
7:42 A.M. EST.
SOMEWHERE IN GEOSYNCHRONOUS ORBIT.

The large satellite began a slow, deliberate turn away from the planet below. Despite being over one thousand miles away, the satellite's tracking computers had the targets locked in. A massive amount of energy surged through the satellite and erupted out of a tube on one end of the satellite. While an observer close to the satellite would have detected nothing, an invisible stream of charged particles left the satellite at close to the speed of light. In under a second, the beam reached its target, vaporizing the raw lunar soil as it passed through the beam.

✛

8:36 A.M. EST.
SPACE STATION GEORGE WRIGHT.

The newly relieved work crews slowly filed onto the same external tank that their replacements, and Jason, had arrived on.

The blare of sirens interrupted the good-byes. Everybody in the room froze and looked up. They knew that something extremely serious had just occurred at the station and were waiting for the announcement that would follow.

"Attention, attention, this is not, I repeat, not a drill. We have just received a major solar flare alert. All personnel are directed to proceed directly to the storm shelter in the Hub. You have approximately twenty minutes to reach the shelter before radiation levels increase to dangerous levels. This report will be repeated at one-minute intervals until the crisis is over."

All over the Wheel, people began to scurry towards the Spokes and the elevators that would take them to the safety of the storm shelter. The Station's designers knew that while the mass of the Station would provide protection from normal levels of radiation, an emergency shelter would be necessary to protect the entire crew of the Station from the dramatically heightened radiation levels that occurred during solar flares. They had constructed such a shelter in the industrial section of the Hub in order to take advantage of the ability to fit more people into a smaller space in zero gravity. While the space would be cramped, most flares generally died down after several hours and the inconvenience would be temporary.

Much smaller but equally effective shelters had been located at the site of the solar power satellites to house the work crews in the event that they couldn't make it back to the Station in time. What none of the people on board the Station other than Jason, Jamie Hampton and Juan Vasquez knew was that the warning was false. No solar flares had been detected. Warnings had not been sent to the crews working at the power satellites and they remained at their stations.

At the time they released the warning, Jason, Jamie and Juan were already on the Hub, hiding in a small closet across from the air lock that led to the only personnel transfer vehicle currently at the Station. They had built just enough time into the warning to enable the saboteur to figure out that he would be trapped on the Station at the time the lunar materials would impact the Station, destroying it. The external tank didn't have a storm shelter built into the design and would not be launched from the Station. They figured that the saboteur would be on the Hub or on the external tank, which was scheduled to leave half an hour before the Station would be destroyed.

Knowing that the Station was about to be destroyed, the saboteur would have to chance the exposure to the radiation of the solar flare and make a run for the nearest power station and safety. Juan had a monitor for the surveillance camera located in the hall approaching the air lock. As a precaution, he had programmed the computer not to open the air lock.

They didn't have long to wait. Juan signaled Jason and Jamie to look at the monitor. They did. A figure entered the hall, but they couldn't make out the person's face. Juan pulled a large pistol out from one of the pockets of his overalls. Jason was about to ask if it was safe to fire a gun on board the Station when Juan stepped forward, opened the door a crack and fired the weapon through the crack. Jason didn't hear a report from the gun or see any smoke, but he did hear a thud as a body hit the deck of the Station immediately outside their door.

Juan vaulted out of the room with Jamie right behind him. Jason, a bit startled, was not so fast. When he got out of the closet, he heard Juan shout: "Son of a bitch!" as Juan delivered a savage kick to the body lying on the deck.

"Why'd you do that?" Jason asked as he walked over to Juan and Jamie looking down at the body clad in blue overalls.

"Because I'm pissed. The bastard had a cyanide capsule in her mouth. She must have bit into it when my tranquilizer dart hit her neck. She's dead. We may never know who was behind this."

"Who was she?" Jason leaned over to look at her face.

It was Jamie that responded. "Nancy Jackson, one of the original crew members. She helped design the mass catcher. I can't believe that she would do a thing like this! After all, she contributed so much to the program. She could have done something at that time."

Juan responded with a sardonic look on his face. "Yes, but she was far too smart for that. Her timing would have been perfect if we hadn't stopped her. She could have caused delays early on by screwing up her design work. But she knew that, eventually, we'd have caught on to her and replaced her. In addition, at that point, political support was solidly behind the program. All she would have done was create a slight delay."

"By acting now, in the manner she did, she might never have been caught and she would have badly crippled, if not destroyed, Whitlock Industries and the solar power satellite program. If she'd waited another year, until we had a portion of the large colony built out at the L-5 point, she would have been too late. The blow would have hurt seriously but would not have been crippling."

Jamie shook her head. "I still don't get it. Why did she do it? How could anybody object strongly enough to what we are trying to accomplish up here to kill over two hundred people?"

"I don't know, Jamie, but I intend to find out. If you will both excuse me, I'll get this mess cleaned up."

"Good. I'll go back up to my office and call off the 'drill' in ten minutes. Will that give you enough time?"

"Yes, it will."

Turning to Jason, "What do you want to do now? Your job is over up here."

"For the time being, but I'll be back eventually." Jason smiled. "I'm planning to emigrate to one of the L-5 cities once they are completed."

Jamie raised one eyebrow. "Really? Glad to hear it. But what about now? If you don't catch the external tank back to Freedom, it will be several days before the next trip down."

"I think that I'd better catch that flight."

"I thought you would. That gives us about an hour. Why don't I give you a proper tour of the Station?"

"I'd be delighted."

CHAPTER 24

The hottest weather in record memory has seared its way across the southern Ukraine, sparking grass and forest fires. Fires in the Crimea are threatening the city of Yalta. Citizens are digging trenches around the city to prevent the spread of fire into the heart of the city.

In contrast, a relatively short distance to the north, in the middle Volga region of Russia, the heaviest snowfall in decades has halted the final harvesting of the fall potato and grain crops and is hindering the shipment of crops already harvested. Government officials estimate that at least fifteen percent of the fall harvest will be lost and that local farmers may be unable to plow for their winter crops, threatening the next harvest as well.

✣

MOSCOW, RUSSIA.

Nicholai Stipanovich Greschko sat behind his large mahogany desk, a happy man. Everything was going his way. His companies had just turned their best profits ever, thanks to the government contracts related to the establishment of the base on Mars. Several European governments clamored to join the bandwagon now that he had established a firm foothold on Mars.

They seemed willing to pay practically whatever he asked for access to his facilities. And even that was cheap, relative to what it would have cost them to get there on their own. Even his most potent competitor, Whitlock Industries, had contracted with him for the launch of several key components on the massive rocket he had developed to launch payloads to Mars. While his attempts to put his competitor out of business had failed, he remained happy. Let them build their space cities and their bases on the Moon. He would own another planet! And he would, ultimately, be in a position to deal with them.

Thanks to the assistance of his personal fourth directorate he had the Russian government and press by the balls. He doubted that even Stalin had wielded the power that he now controlled. But he was not satisfied. He plotted and planned to have power unmatched by any leader in history.

He had gained a foothold in the Congress of the United States. While he had managed to use his influence to defeat President Tom Hammond, a man he had feared, he was frustrated in not having his man elected. Unfortunately, even his vast power could not accomplish everything. His influence had to remain invisible in America. He must never let the Congressmen he supported learn of the strings that ultimately led back into his hands.

He stood up and walked over to a bucket of caviar that sat in a large Baccarat crystal bowl. After several mouthfuls, he noticed that he had dropped a few specks of the caviar on his trousers. He smiled to himself and pushed a button on his desk. In walked his personal assistant, Ekaterina Golganov, the twenty-three year old daughter of his former adversary. He watched her closely as she walked across the room toward him, her breasts bouncing against the tight silk blouse he had purchased for her in Paris. He had come across her several years earlier, at the

funeral for her father. After several subsequent 'chance' encoun-
ters he had invited her up to his office for an interview. She had
proven herself to be willing and eager to do what Greschko
required of her. Unlike her father, she knew the way the wind
was blowing in the new Russia. He pointed to the caviar in his
lap. She knelt down in front of him as he smiled.

A cold wind blasted Ekaterina as she exited the Metro station.
Despite the fact that she wore the mink coat that Greschko had
given her, she did not wrap herself more tightly in it. In fact she
wore it very loosely, as if it were something repugnant. She
walked down several streets to the apartment where she and her
mother lived. Despite being Greschko's mistress, she did not
have her own car or live in one of the new fancy apartment
buildings that one of Greschko's companies had constructed in
the heart of Moscow. She refused to move into the apartment
where they frequently spent evenings. She insisted that her
mother was ill and needed her. Given what she did for him
when they were together, he did not insist.

When she arrived at home she flung the coat to the floor and
walked into the small living room where her mother was laying
on the sofa. Their neighbor departed after Ekaterina arrived. She
sat down beside her mother and began to stroke her hair. Her
mother no longer recognized her. Ekaterina began to cry, thinking back about the life they had led before Greschko had killed
her father, an influential cabinet minister, driving her mother
over the edge.

CAMP DAVID, MARYLAND.

In what the press called a magnanimous gesture, President Tom
Hammond had invited President-elect Geoff Barnes to Camp
David for the weekend to discuss a smooth transition of power.

What the press did not realize was that the President greatly preferred the President-elect to his own party's candidate. Hammond had been a lame duck since his failure to capture his party's nomination at the party convention. The liberal wing of his party, led by George Gallager, had cast him out.

Now, he would hand power over to the man he had secretly supported all along. In the meetings with the other heads of Project Runaway, including Jason Graham and his wife Samantha Graham, Tom Hammond announced his intention to move to the first space city upon its completion later the following year. The entire committee agreed that the move would be announced shortly after the opening of the space city.

Samantha's father would announce that President Hammond had agreed to become a direct part of the bold experiment that he had started by becoming the first mayor of the first space city. While Whitlock Industries would own the first city, they planned to eventually sell ownership of the city to its inhabitants. President Hammond would take the lead in setting up a government for the city.

Unfortunately the signs of global warming were becoming more visible. While Jack Watson and the others at EnviroCon had managed to disseminate a significant amount of false and misleading information, the popular press had begun to speculate again. The members of the committee decided that it was time to bring several more people into the still relatively small group of people that knew about the crisis. All of the members of the committee, led by Jeff Nakamura and Arthur Jacobson, had compiled a list of leading scientists that they would consider approaching with the information about the runaway greenhouse effect.

Hamilton Smith had compiled dossiers on all of the scientists listed. He had summarily deleted a number of the names based upon security considerations. He had also designated several of the candidates as 'prime' as a result of information he had about them that would ensure their cooperation. While most of the members of the committee agreed to not consider such criteria in selecting names to approach, most of Hamilton Smith's names made it to the final list. Each of the listed scientists would be carefully approached by members of the team in an attempt to convince them to either actively publish false conclusions in an attempt to hide the truth from the public or, at a minimum, keep whatever conclusions they might reach about global warming secret. If they could not be convinced, both the President and the President-elect had co-authored an urgent appeal for support. While the plea from a current and a former President should be persuasive, the letter also contained the ultimate persuasive measure, a guaranty of a place on one of the space cities for the scientist and his or her family. Finally, if stubborn scientific integrity still controlled the individual's decision, both Presidents, current and future, had executed an executive order declaring the information a matter of urgent national security and top secret. A revelation of the information would be deemed an act of treason, subject to life imprisonment.

While risky, if they could convince several prominent scientists to support the spurious data being disseminated and to ridicule those championing global warming, they should be able to delay the day of reckoning when the public at large would perceive the threat as real.

Unfortunately, the fires created by the Saudi - Iranian conflict and the eruption of several volcanoes had accelerated the process. The computer models now predicted that the ice cap over the Arctic Ocean, which had already begun to shrink, would disappear entirely within the next five years. While the melting of the ice would not result in an increase in sea level — the North Polar Ice Cap was sea ice as opposed to the far larger ice cap which rested on Antarctica, a land mass larger than Australia — such an event would be very difficult to explain away.

The Camp David meeting concluded on a positive note. Even though they would be cutting things rather close, the computer predicted that they would have established a viable and independent civilization at the Earth's Lagrange points before support from the Earth was cut off.

✦

McLean, Virginia.

Arthur sat at the computer terminal in his office at EnviroCon, his head down, his hands pulling at his hair. This was the last straw. He'd just gotten Nakamura's memo about having to silence several friends of his that had begun to speculate about anomalies in the NASA data. Yes, they would try to persuade them first, but Arthur knew they'd go farther than that. Knowing that he was working with EnviroCon on the project, an old grad school friend of Arthur's had called him about the data. It had just been released. Many scientists, particularly those with ties to Big Oil and Big Auto, had jumped at the data, using it to explain away fears of global warming. The press picked up on those stories and banished reports about global warming to the tabloids. But good scientists were not fooled too long. Arthur's friend knew. He had studied other data and picked up several anomalies that could not be explained away. He called Arthur for his opinion on the anomalies. Arthur had looked into the anomalies and discovered several things he hadn't expected. Much of the anomalous data predated the disinformation campaign. In fact, many of them were the same anomalies that Jason's program had dismissed. Arthur had dug a little deeper and uncovered that most of the changes had Jeff Nakamura's mark all over them. He called his friend back to try to explain away the anomalies, but the phone was answered by the friend's widow.

Arthur didn't know who to trust. At the very least, Jeff Nakamura was involved with a group that had been trying to hide global warming for far longer than the Project Runaway team had known about it. At the worst, his friend was killed as a result of his call to Arthur. Arthur had decided to approach Jason Graham. He knew that Jason couldn't be involved in the cover up that had taken place before Project Runaway. He tried e-mailing his concerns to Jason, but it backfired. He'd never forget the day that Jeff Nakamura and Hamilton Smith entered his office. They sat down, Jeff lit his pipe and they dropped printouts of his e-mails on his desk. Before he could recover, they shoved pictures from the scene of his friend's "accident" in front of him and then pictures of his children at college. They didn't say a word; they didn't have to. He got the message. He was unwilling to risk the lives of his children. He knew he was trapped. He would have to go along with them and continue to help them. They knew that he would be persuasive with his friends. And he would be. He would go along with them. He had no choice.

CHAPTER 25

The meteorologist aboard the Space Station Freedom signaled the National Hurricane Center in Miami that a new low-pressure cell had begun to build in the Gulf of Mexico. While the United States had been lucky for several years and had not suffered the destruction of another Marla, conditions were ripe for a massive storm to build over the Gulf. Water temperatures remained extremely high and current wind patterns supported the development of a major storm.

After watching the storm for twenty-four hours, the meteorologist on board the Freedom as well as his counterparts at the National Hurricane Center knew they had trouble. A strong eye had formed and winds were rapidly building up. Computer projections predicted that the storm would develop winds in excess of 200 miles per hour and that there was a 90% chance that the storm would impact the Gulf Coast.

At 8:45 p.m. Emanuel Ortega, the head of the National Hurricane Center, picked up the phone and dialed the new President. It surprised Ortega that it only took him forty-five minutes of arguing with members of the President's staff and the National Science Advisor to get through to the President.

"This is President Barnes speaking."

"Mr. President, my name is Emanuel Ortega,. . ."

"I know who you are, Mr. Ortega; and I know we are facing a grave situation. Please come to the point."

"Mr. President, as I am sure you are well aware, Hurricane Gwen is shaping up to be the most disastrous storm to hit the country since Marla. It could be worse if it hits the heavily populated lowlands of Louisiana."

"I know that. I've seen copies of your reports. What do you advise that I do: order an evacuation of the entire Gulf Coast from Brownsville, Texas, to Pensacola, Florida? Do you have any idea what that would entail?"

"Yes, Mr. President. I do. But we have developed an alternative that we think will work."

"You think it will work or it will work."

"Sir, in my opinion, it will work. Ever since Marla, we've been investigating ways to lessen the strength of hurricanes. We believe that we have a way to do it, but we need your authorization."

"Go on. I'm listening."

"Our computer models predict that if we simultaneously detonate two small nuclear devices on opposite sides of the eye wall, the blast will destroy the storm's organization and its strength. With cloud seeding to follow, the storm will become no more than a large, disorganized low."

"Are you proposing that I authorize the detonation of two nuclear devices?"

"Yes, Mr. President. It will work."

"How long do we have to implement your plan?"

"Ten, maybe twelve hours before the eye gets too close to land."

"I'll get back to you." The President hung up the phone. Geoff Barnes knew that his Presidency would be extremely difficult, but he hadn't anticipated making such a tough decision this early in his Presidency. He picked up the phone.

Several hours later, he made his decision. His advisors confirmed that the type of device needed was very clean and would not create any significant risks of fallout. While he had managed to speak with several key world leaders, he decided that he had to take action before he could obtain the consent from more than a handful. Decisive action was called for and he took it. After signing the requisite orders directing the Air Force to coordinate with the National Hurricane Center, he went on national television to announce the decision knowing that if he had made the wrong decision, it might cripple his Presidency.

The President succinctly presented the population with the nature of the crisis facing the Gulf Coast. He outlined the risks associated with a massive evacuation and the need for an alternative. He then announced the alternative, that the United States would use several small nuclear devices to subdue the hurricane and eliminate the necessity of the costly evacuation and the damage that would result if the storm made landfall as projected. The pictures of what Marla had done to Palm Beach served as a graphic reminder of the risk being faced.

After the speech, the President returned to the Oval Office to watch the results of his action. Cameras aboard the Space Station Freedom beamed live pictures into living rooms all over the globe as the two Air Force jets released their cargoes over the hurricane and rapidly pulled away. The bombs detonated as planned. A bright flash erupted in the eye of the hurricane. As winds in the upper atmosphere dispersed the mushroom clouds, the cameras revealed that the eye of the storm had disappeared. The plan had worked.

When what had once been Hurricane Gwen came ashore, it did so as a disorganized front. While it dumped torrential rains and spawned numerous tornadoes, the anticipated destruction did not happen. Once it was shown that no fallout had reached land, the President's popularity skyrocketed. A method had been developed to halt the threat of massive storms wreaking havoc along the coastal regions of the country. The President and all of the members of Project Runaway breathed a major sigh of relief.

✦

JUNE 11, 2013.
LOS ANGELES, CALIFORNIA.

After weeks of stifling heat and months of drought, the death of a Mexican gang leader at the hand of a rival black gang has triggered the riots feared by governmental officials for years. With the situation rapidly deteriorating, hundreds are already dead.

One of the rival gangs managed to blow up a power substation, cutting off electrical power to hundreds of thousands of residents. Due to the lack of electricity and open fire hydrants, the pumping stations for water are only able to deliver a trickle of water to hundreds of thousands of residents. Given the searing hundred degree plus heat, the riot-related fires raging out of control, and the lack of water and electricity for air conditioning, health officials fear that thousands of the sick and the elderly will die within the next twenty-four hours.

While the Mayor and the Governor have declared martial law, the National Guard troops ordered into the area have been turned back by automatic weapons' fire. Instead of restoring order the National Guard has cordoned off sections of East L.A. and Watts to contain the rioting and looting. News helicopters reported that the gangs appear to have begun an all-out war on each other while the riots continue; however, after several shots penetrated one news chopper, seriously wounding the cameraman, the Mayor has ordered all news helicopters grounded.

The President, at the request of both the Mayor and the Governor, has ordered the Army into the area. It is hoped that the mere presence of the troops, supported by tanks, armored personnel carriers and helicopter gunships, will quell the rioting without having to utilize their firepower. But the threat remains a real one. Under the Riot Control Act of 2012, enacted by Congress after the Miami and Chicago riots last year, the troops are authorized to use deadly force to restore order.

The heat wave and the drought have extended beyond California into the Great Plains. The loss of corn and wheat crops is estimated to be extremely heavy. Projections are that the crop this year will be forty-five to fifty percent below last year's moderate crop. The President is considering selling the government's grain reserves on the open market to prevent shortages and price gouging. After extreme public outcry, the President terminated the humanitarian grain shipments to Central Africa to provide for those suffering hardship at home.

A positive note in the bleak news today. Former President Tom Hammond will become the first former head of state to leave the bounds of Earth. After a brief public statement Mr. Hammond will ride atop the privately owned Von Braun space liner on the first leg of his journey to be the guest of honor at the christening of Clarksville, the new space city built by Whitlock Industries. While President Barnes was scheduled to attend Mr. Hammond's launch, due to the crisis in Los Angeles, he has canceled his trip.

✦

JULY 8, 2013.
CHACO MESA, ARIZONA.

Jason hardly recognized Chaco Mesa as he flew over it. The rectenna field still dominated the landscape, but only because it had been expanded to almost twice its original size in order to accommodate several more solar power satellites. To one side of the rectenna field, Jason could see the partially completed Chaco Mesa Arcology. While compared to the rectenna field it appeared small, once completed, it would be the equivalent of a small city within a single structure. Over ten thousand people would live, work, play, go to school and shop within the structure. At forty stories tall and half a mile long, it would become the world's largest building.

He felt a tug at his shirtsleeve as the plane circled the air/space field. "Daddy, is that a spaceship?" Jason looked out the window to where his son was pointing. He could see the Von Braun sitting at the spaceport end of the terminal building.

"Yes, son; that's the Von Braun." Little Jason sat up higher in his seat to look out the window. He pressed his face against the window.

"Is it about to take off? It looks like smoke's coming out the back of it."

Jason looked over his son's shoulder, a bit concerned at the mention of smoke. He chuckled to himself. "No, son, it isn't. It won't take off for several hours yet. That 'smoke' you see is actually frost. They are fueling the ship. The fuel is so cold that the water vapor in the air freezes and turns into a cloud of frost." He wasn't sure whether little Jason would understand his explanation, but his son nodded as if he did, far too proud to admit it if he didn't understand. Jason looked over at Sam and noticed her smiling at he and his son. He patted her leg and smiled back at her.

The stewardess wandered by and had to ask little Jason to sit down and fasten his seat belt for landing. He did, with a little fussing, and extended his hand for his father to hold as the plane banked for its final approach. After landing the captain announced that they would be taxiing past the Von Braun and that passengers on the left-hand side of the aircraft would have a great view of the spaceship. Little Jason fidgeted in his seat with excitement. They slowly passed by the Von Braun and little Jason stared out the window with his mouth agape.

Jason leaned over toward him. "You know, Jason, next year you and your mother are going to fly on a spacecraft just like the Von Braun."

"Really? No kidding?" Little Jason looked up at his mother to receive confirmation of the news.

"No kidding sweetheart. We'll be riding up next year to spend the whole summer on Clarksville."

"Oh boy! I can't wait to tell my friends. Boy, will they be jealous."

A special liaison officer for the company met them at the gate. He led them through a door labeled 'Employees Only' down a set of stairs to a waiting cart. "Your bags will be picked up and dropped off at the hotel in the Arcology later. Right now

I'm taking you directly to the operations center."

After a brief ride in the cart, they came to a large air lock. Jason recognized it as an entrance to the magnetic rail line. They only had to wait for a few minutes for a train to arrive. When it did, the air lock opened and they stepped into the train car. Their guide told them that it would not take off until they fastened their seat belts. The heavily-cushioned seats were built into the sides of the car, facing the center. Once they were seated and after Sam had carefully checked little Jason's seat belt, the train began to accelerate. As it did, the seats swiveled so that they faced forward.

After only a few minutes, the acceleration slowed and then stopped. The seats swiveled so that they all faced backwards. Deceleration, or Jason thought what might better be termed braking, began in the opposite direction. They decelerated for the same amount of time as they had accelerated and the train slowly came to a halt. The seats swiveled back to face the center of the car and everybody unbuckled and got off. As they exited, their guide asked little Jason how far he thought they had traveled. He said a mile. The guide shook his head and said that they had actually traveled over twenty miles under the rectenna field to the private, underground operations center of Chaco Mesa. Little Jason nodded his head as if that was not a big deal. Their guide led them up another short stairway into the operations center.

As arranged, the guide led them first to a small room filled with screens and computer terminals and virtuality devices. Sam leaned over. "Jason, would you like Mr. Samson to show you what it would be like to take off in the spaceship today while your daddy and I go to a short meeting?"

Little Jason barely hesitated. The room put the video arcade at the mall to shame. He nodded vigorously and headed for a virtuality headset. Mr. Samson walked over to help him. Jason and Sam left the room and walked into the small conference room next door. Former President Tom Hammond and Jack Watson were both in the room with Dr. Gordon Templeton. All three heads turned as they entered. Jack walked over to greet them. He gave Sam a hug and a kiss on the cheek and then shook Jason's hand.

"Sam, Jason, it's great to see both of you again."

"Nice to see you again, Jack." Jason was surprised at Jack's appearance. His eyes were swollen and he looked twenty pounds heavier than the last time Jason had seen him. Even worse, his breath smelled like scotch. "How's Michelle doing?"

Jack looked down toward the floor momentarily and then back at Jason. "We're separated."

"I'm really sorry to hear that, Jack. If there's anything either of us can do, please let us know."

"Thank you both. But I doubt there's anything anyone can do. Anyway, that's not what we're here for. Come on, President Hammond wanted to see both of you before his speech." Jack led them over to where the former President and Dr. Templeton were in the midst of an animated discussion. They stopped when Jack arrived with Jason and Samantha.

Tom Hammond turned to greet both of them. "Samantha, Jason, wonderful to see you."

"Nice to see you again, too, Tom, although I'm afraid our visit will be rather short."

Tom looked at his watch. "You're right, Jason. I've got an appointment with the press, hopefully the last one I'll have to endure for quite sometime."

Sam smiled at the comment. "The last one you'll have to endure in person. As Mayor of Clarksville you'll need to endure additional press conferences."

"Just wishful thinking at that, then. Oh, well, at least when I'm in Clarksville, I'll be able to duck the hard questions by claiming I didn't hear them due to static." Everyone laughed. "By the way, I didn't mean to be rude. This is Dr. Gordon Templeton. Dr. Templeton, Samantha and Jason Graham."

"I already know Jason, but I haven't had the pleasure of meeting his lovely wife." Dr. Templeton gave a slight bow to Samantha.

"You have accomplished a great deal since we first met,

Gordon. I hardly recognize the place."

"In fact, that's one of the things that I was discussing with President Hammond when you arrived. He questioned the need for all of the additional facilities that we constructed here. He feels that we should spend the money building more facilities in space rather than here at Chaco Mesa and at the Colorado, Washington State and Georgia sites. I disagree. As glorious as the space cities are, they will always remain secondary to the facilities here on the planet."

"Doctor, let's not reargue the point. Both facilities have their purposes." The former President glanced over at Jason and Jack Watson. While Dr. Templeton was not privy to the reason why the solar power satellites and space cities were being built, the members of the Project Runaway executive committee had argued the same point months before. After seeing the Chaco Mesa facility, several members of the executive committee, including Jack Watson, had argued in favor of building more Earth-bound arcologies and building fewer space cities.

Clearly more arcologies holding more people could be built in the time they had with fewer resources, but Jason and the computer models continued to favor the development of space cities. The former President agreed. Ultimately, their hope remained the establishment of a viable civilization off of the planet. While the arcologies could now be built to survive the environmental consequences of the runaway greenhouse effect, at least as far out as the computer models could accurately predict, the political and sociological implications remained problematic. Jason and the former President pointed out over and over again that the position of the arcologies was not tenable. Once the millions of those condemned to live outside of the arcologies learned of their existence, they would fight to gain admittance.

President Barnes continued to support the push for space. But the Vice-President, Joel Friedman, vigorously argued in favor of the arcologies. With the Black Berets to defend them and the technological base being built around them, as had happened at Chaco Mesa, he argued for the abandonment of the space settlements altogether.

Fortunately, in Jason's opinion, the voice of the Vice-President remained in the minority. In fact, Jason now regretted the executive committee's decision to inform the Vice-President of the existence of Project Runaway.

Dr. Templeton was paged and he excused himself. After Dr. Templeton left the room, Jason turned to the former president. "Why did you even start the discussion, Tom? What was your point?"

Tom Hammond glanced at each of the individuals in the room. With a nod from Jack Watson, he began. "I wanted to find out whose side he's on. In light of our conversation, I doubt he's been approached."

"What do you mean, 'approached'?" Both Jason and Sam looked surprised. Jack Watson interrupted. "Jason, while you are a brilliant man, one of your weaknesses has always been in reading people. You are far too trusting. Tom and I have our suspicions as to the way things have been going within the Project. The Vice-President is establishing his own power base within the organization."

"I still don't get it."

Sam turned towards Jason. "I think that what they're getting at is that your fears about the Project are going to be realized. Is that correct, gentlemen?"

"Basically, yes. Listen, Jason. I've only got a little while before the press conference. But I wanted to take the time to warn you and Samantha. I'll secure the high ground for us. But be leery of some of the Project members' motives; the Vice-President and General Maxwell in particular. I fear that when push comes to shove they may try to gain control of the Project and the nation."

"But what about Geoff Barnes? I can't believe that he'd put up with anything like that. I've known him far too long."

"It's not the President that I fear. Instead, I fear for him. Not all of the Project's funds have gone where they were supposed to."

"Why not confront them now with our suspicions, while we can?"

"Because we still need everybody to work towards our ultimate goal, the survival of mankind. As long as we have the support of President Barnes and the votes on the executive committee, they'll go along with us and support us. If we confront them, we may create a rift that could topple the entire Project. That's a risk that we cannot take."

"Jack has agreed to support the Vice-President's group in committee meetings."

"That's right. While I still fully support your original proposal, Jason, I've begun to voice support for the Vice-President's alternatives. It's my hope that they will confide in me and draw me into whatever plans they might have."

"I hope we're being paranoid. But for the time being, we wanted to warn both of you. Don't trust anybody on the committee." Dr. Templeton reentered the room.

"I apologize for the interruption. Some fool technician got a message screwed up for me."

Tom Hammond stepped forward. "That's quite all right, Doctor. We were just finishing up." The former president turned toward Samantha first. "Samantha, I guess that the next time I'll see you, you'll be visiting my fair city." He bent down and kissed Sam's cheek.

"Little Jason and I are both looking forward to the trip."

"Jack, I hope that you can make it up at some point as well."

"I'm sure that I will, Tom." Finally the former President turned towards Jason.

"Jason, I've enjoyed working with you and hope that you can come up for a visit as well. I'll give you the key to the city. Although, no doubt, you've already designed your own key in the computer system." They all laughed.

Jason shook Tom Hammond's hand. "Have a good trip, Tom. I'll see you next summer if I can get away with Sam and little Jason."

Sam and Jason picked up little Jason and went over to the hotel in the Arcology to change before takeoff. They traveled via the underground rail system to the air/spaceport and watched from the VIP viewing lounge. After the takeoff, they began their vacation in earnest. They spent several days hiking, fishing and horseback riding in the wilderness areas that surrounded Chaco Mesa. It was the first real vacation that Jason had taken with his wife and son for several years. He and Sam managed to forget about the Project for a couple of days.

CHAPTER 26

The longest drought in Sub-Saharan Africa in over a century has combined with searing heat to create the greatest famine in over a century. Despite pleas to the United Nations and the world, help has not been forthcoming. While the countries of sub-Saharan Africa are screaming for relief, it is difficult to blame the West for its lack of support. The United States is in its third year of record small harvests due to unusual weather conditions attributed to the massive El Niño event that has gripped the Pacific for the past four years. Reserves in the U.S. are running low. The President has indicated that the government cannot contribute food to the relief effort but has made a direct plea to the people to contribute.

Europe is not in much better shape due to flooding in the farm belts of France and Germany. Some scientists are attributing the unusual weather conditions to a weakening of the Gulf Stream and to the diminishing size of the Northern Polar Ice Cap. Only the Ukraine has managed to send a sizable amount of food to starving Africa after their second record harvest in a row. However, such relief is only a drop in the bucket. Hundreds of thousands are projected to die each month due to starvation, malnutrition and related diseases.

Animal populations have fared even worse. Despite the efforts of the government of Kenya to protect its elephant herds, scores of desiccated elephants were found near a dried-up watering hole in the largest game preserve in the country. Other elephant herds as well as antelope, zebra and other wild species are being hunted mercilessly by roving bands of armed men. The governments of Chad, Uganda, Central African Republic, Ethiopia and Somalia have already disintegrated. Chaos is rampant in Africa. It is feared that several other nations, including the Sudan, Kenya and Nigeria could succumb to the chaos if conditions do not improve by the end of the summer.

In related developments, the Reverend Hosea Johnson has brought his crusade with its thousands of followers to Washington D.C. to preach the futility of fighting against the forces of God. The preacher, whose following has grown rapidly over the past year, is predicting that the rapture is upon us. People should stop working and give up their worldly goods to prepare for the end and the coming of the Lord. Between the turmoil in Africa, the droughts and crop failures in this country and the crop failures in Europe, it is understandable why the Reverend's following has grown. The four horsemen are indeed loose upon the globe.

✦

WASHINGTON, D.C.

President Barnes looked out of the window of the Oval Office. He could see the protestors gathered out in front of the White House. They had become a constant presence over the past year.

The President sat down at his desk, pulled out a bottle of aspirin and a glass of milk. He swallowed the aspirin to relieve a headache that never seemed far away. His first appointment was with the Kenyan ambassador. He knew what the ambassador would ask for and knew that he would turn him away unsatisfied. As he reviewed his notes for the meeting, the telephone rang.

"Yes, Teresa."

"Mr. President, I've got an urgent call from Mr. Jason Graham. He says it's imperative that he speak with you immediately."

"I'll take the call. This may take some time. Get the Secretary of State to express my apologies to the ambassador and reschedule the appointment for me."

"Yes, sir."

President Barnes picked up the phone. "Good morning, Jason. How are you?"

"As well as anyone can be these days, Mr. President. How are you?"

"I think you phrased it nicely. Now, what can I do for you?"

"Mr. President, we have a green light on our boards. Can you make the necessary arrangements?"

The President caught his breath. A green light meant that a major crisis had occurred with Project Runaway. What it was, the President could not imagine. All he knew was that Jason needed to talk to him immediately and could not discuss it over the White House lines. Green light meant meet at Camp David as soon as possible. "Yes, I can arrange that."

"Good. Thank you, Mr. President."

The President hung up the phone and buzzed his administrative assistant. Teresa entered the Oval Office immediately. The contrast between her attractive outfit and her hideous shoes would have usually made the President chuckle, but not today. "Teresa, please get my car ready. I'll be going out one of the back ways. I need to get up to Camp David immediately. And, by the way, get the National Security Advisor to assemble a crisis evaluation team." The President looked down at his watch. "Have them, in the situation room at 3:00. Oh, and cancel all of my appointments today."

"All of them, Mr. President?" Teresa raised one eyebrow.

"Yes, Teresa. I know what a stink it'll create, but make up an excuse. Say I've come down with the flu."

"Very good, sir."

Just over an hour later, the President, Jason Graham and Jack Watson met in the President's private office at Camp David. He didn't like the look of his two friends. With bags under his eyes, Jason looked exhausted. Jack looked even worse. His hands shook slightly in the President's grip. Although, thinking to himself, he doubted that he looked much better. His three years in office had aged him. If not for his makeup team, he'd look like a man twenty years his senior.

"Gentlemen, let's get to the point and quickly. Based upon your green light alert, I called together a meeting of my crisis management team for 3:00 this afternoon."

"Good. I'm glad that you did. I'm afraid that we have finally reached the point of no return with the Project. I hate to say it, but, given the way things were progressing, this may be a blessing in disguise."

"What do you mean, Jason?"

"Something has come up that gives us both little choice and the perfect opportunity to go public with the runaway greenhouse effect. Geoff, satellites have shown a series of shifts in several of the largest Antarctic ice fields over the past several weeks. Our computer models indicate that several massive sections of the Ronne Ice Shelf will break off into the Weddell Sea."

"So, the northern ice cap is already almost gone. Didn't we anticipate the breakup of the Antarctic Ice Pack as well?"

"Not really. Actually, we thought that the volume of ice on Antarctica would increase due to an increase in the amount of water vapor in the atmosphere caused by higher temperatures. What we've recently discovered is that a slight rise in sea level has created a great deal of instability in the ice pack."

Jack interrupted. "Get to the bottom line, Jason."

"Mr. President," Jason paused. "Geoff, when the massive section of the ice shelf breaks off into the Weddell Sea, it will shift so much weight that entire ice fields will shift. It will be the equivalent of an 9.1 level earthquake, creating a monster of a tsunami. Our models show that every major port city on the Atlantic will suffer significant damage. If we do not evacuate those cities, tens of millions of people will die."

Silence pervaded the room. After several minutes, the President again spoke. "And you are certain of your results?"

"Yes, Mr. President."

"How much time do we have?"

"Not much, 96 hours at the outside."

"Is there anything that we can do to stop this catastrophe?"

"No, sir; not that we could come up with in that short a time."

"What do you propose?" Jack took over for Jason.

"Geoff, when Jason shared his preliminary results with me, I focused my team on coming up with a strategy to cope with the crisis. You've got to declare a national state of emergency and impose martial law."

"You know they'll impeach me if you're wrong."

"We know that."

"I'll also need the support of the military and my crisis management team. I'll need both of you in the White House Situation Room at 3:00 for the meeting."

"No problem. We have reams of data and projections to support us. In addition, both the Vice-President and the National Science Advisor know the truth about Project Runaway. They'll support us."

"Granted. With their support, we should be able to convince the others. But then what? Assuming that we can move tens of millions of people in the four days you say we've got, what do we do with them?"

"Geoff, our team feels that this is the ideal time for you to take control of the government. After the initial crisis is over, you will simply not lift martial law. At first you can blame it on the need to relocate and feed the refugees. That will buy you several weeks, if not months. During that period, we can leak information that will confirm that the real problem is global warming. You will suspend civil rights and delay the election indefinitely. You will, in essence, become the last president of the United States of America."

"I knew that Project Runaway would lead to this eventually, but I had hoped that I'd be out of office by then and able to join Tom Hammond on Clarksville. This is a step I'm not sure that I'm ready to take."

Jason walked over to the President and put his hands on his shoulders. "Geoff, we've known each other for a long time now. I've watched you handle the numerous crises that we've faced before and during your Presidency. I can think of no one else that I would want to lead our country through this crisis. I know that you will not abuse your power and authority. Besides, you've got to do it. If you don't, who will?"

"Jason, Jack, I appreciate your confidence and I know the arguments. But I need a little time."

"A little time is all we have left, Mr. President."

"I know. I know. I'll see both of you at 3:00. Now, please, I need some time alone." Jason and Jack both nodded and left the room. They knew what Geoff Barnes had to do. They did not envy him.

The President left Camp David and returned to Washington, D.C. He did not go to the White House. Instead, he drove to the Lincoln Memorial. He no longer cared what the press would make of his being seen out in public after he had released a statement indicating that he was ill. It wouldn't matter what the press said forty-eight hours from now.

His limousine pulled up in front of the Lincoln Memorial. He got out of the car, flanked by his secret service agents, and made the long walk up the great marble steps to stand before Mr. Lincoln. He stood unmoving, but not unmoved, until Jim, the chief of his Secret Service team, tapped him on the shoulder.

"Mr. President, we have to leave. The Reverend Hosea Johnson and his followers were demonstrating in front of the Capitol. They've learned that you are here and are on the way over. We've got to get you out of here or we won't be able to protect you."

After a final look at Mr. Lincoln, the President turned to leave. "I've got one more stop to make, Jim. Can you get me into the National Archives?"

Jim spoke into the microphone built into his collar. "No problem, sir."

"Thank you, Jim."

The President climbed into the back seat of his limo without noticing the crowds of people heading towards them as they pulled away. When the limo stopped, he stepped out and followed the lead of his security team. They led him down a corridor where they were met by the curator of the National Archives. Without even waiting for an introduction, he asked the curator to take him to the Declaration of Independence. It was below the public exhibition hall for routine maintenance. The curator led the President to the room where its display case was being cleaned.

After reaching the room, the President asked that he be left alone. The curator began to object but was rapidly ushered out by the Secret Service team. Jim had served with the Secret Service for most of his life. He had been assigned to five separate Presidents. He had seen other Presidents in moments of crisis and knew it was time to leave the President alone to his thoughts. The President spent just over an hour in the room with the august document before signaling his security team to return.

The change in the President's demeanor pleased Jim. Rather than appearing surly and haggard, the President was himself again, his jaw set and determined. He addressed himself to the curator.

"When is the Declaration and the Constitution due to be returned for public display?"

"Tomorrow, Mr. President."

"I want you to delay that please."

"But..."

"That is a direct order of your President. It is not subject to question. If you like, I'll send a written statement over to you."

"That will not be necessary, sir."

"One more thing, I read somewhere that the Declaration and the Constitution are both stored in vaults sunken well below floor level. Is that correct?"

"Yes, it is, Mr. President. In fact the cases are currently within the chamber that leads down to the vaults. We store the documents there, as well as our copy of the Magna Carta, whenever the Archives are closed. The vaults were designed to withstand anything but a direct nuclear blast."

"That's what I thought. But, tell me, are they watertight?" The curator looked puzzled by the question.

"I've never thought about it, but they must be because they're airtight."

"Good. Keep them in the vault until further notice." The curator nodded. Ten minutes later the last elected President of the United States entered the White House for the last time. As he walked into the situation room to face the greatest crisis in the country's history, one he was not sure the country would survive, he thought of the oath he'd taken when he was sworn in and the silent pledge he had just made to Mr. Lincoln and to those who had preceded him. He would do all he could to return the country to the fundamental principles laid out in the Declaration of Independence as soon as possible.

CHAPTER 27

The breakup of the ice began slowly; first only small sections of the shelves, the size of large icebergs, calved off. As the smaller bergs broke off, cracks in the shelf began to expand and separate. The level of the entire ice shelf, which had already risen over a foot due to the rise in sea level, rose further as the calving of the bergs reduced the weight of the ice shelf. The ice sheet rose off of the sections of land that had supported it for thousands of years. As it did, the shelf shifted forward, rapidly.

The first major section of the ice to go, while large, was not the major catastrophe predicted. As the Larsen ice sheet slipped forward, a five-mile wide and thirty-mile long section of the ice weighing several million tons broke off into the Weddell Sea, creating a massive swell. The swell expanded rapidly and moved away from the point of impact with the water at several hundred miles per hour. A portion of it headed out into open ocean and a portion headed into the Weddell Sea, towards the much larger Ronne Ice Shelf.

Seismographs at the various scientific stations located on Antarctica and in South America and South Africa registered the initial shifting of the Larsen ice shelf and the calving of the first huge berg as an event measuring 6.9 on the Richter scale. Reports were immediately broadcast to locations around the globe.

✦

10:07 A.M., EST.
DENVER, COLORADO.

An Air Force major tore the report off of the printer in the situation room located in the New White House on the outskirts of Denver, Colorado, and handed it to the President.

"Gentlemen, may I have your attention? We have just received our first report from our outpost at McMurdo. They have registered a seismological occurrence of 6.9 on the Richter scale. Charlie, can you pinpoint the location for us on the display screen?"

"Yes, sir. It will be coming up in a minute."

All eyes in the room focused on the twenty-foot by ten-foot display screen located on the wall to the left of the President's chair. Colonel Charles Hanson, the President's senior liaison officer with the Pentagon, worked the computer terminal in front of him.

"I've indicated the epicenter with the flashing red dot, sir."

Jason mumbled under his breath. Evidently, the President heard him.

"What was that, Mr. Graham?"

"Mr. President, that's in the middle of the Larsen Ice Shelf. It looks as if the breakup of the shelves has begun."

"Any analysis yet, Mr. Graham?"

"No, sir. I'd like to feed the data into the computer and rerun the models with the new information."

"Does anybody have an assessment of the event? I know the data is sketchy, but please give me your thoughts."

Captain Scott Purdie of the Navy's oceanographic and meteorological group cleared his throat.

"Mr. President."

"Yes, Captain Purdie. Please feel free to speak up. That's what we are here for, to listen to everybody's views and make an assessment based upon those views."

Captain Purdie glanced around the room focusing on Jason. Jason knew the look; just who the hell was he and who did he think he was. To them, Jason was an upstart. Few in the room knew of his relationship with the President. They knew that he had notified the President of the impending crisis and that he had the President's ear. Most knew their ranking in the pecking order; they didn't know his. "While a 6.9 reading is large, it's not a cataclysmic event." Captain Purdie walked over to the screen with a pointer. "I estimate that the event represents a chunk of ice of approximately 15 to 30 square miles dropping into the Weddell Sea. From my experience with tsunamis in the Pacific, I would estimate that, while a sizable and dangerous wave was created, it will not impact the U.S. mainland. The wave will only be several feet high in the open ocean, but will build up as it approaches the coast of Africa. The wave will, doubtlessly, produce some damage along the west coast of Africa, South Africa and Namibia in particular, it should not be large enough to create a bounce and a secondary wave heading toward the coastlines of North and South America."

"Any response or other comments?" The President looked towards Jack and Jason.

Jack looked over at Jason, "I'll defer to Mr. Graham's expertise, Mr. President."

"Mr. President, again, I prefer to delay any comment until I complete my analysis. I will have something for you momentarily."

"Anybody else?"

Secretary of State Elizabeth Savage stood up. "Mr. President, you know my views. I opposed the declaration of martial law from the beginning. You are extremely unpopular as it stands. The only thing preventing your impeachment right now is shock. If this threat proves false, you will not survive. None of us will. Impeachment will be the last of your worries. As you know, well over one thousand people lost their lives in the evacuation."

"I know, Elizabeth. I know. We discussed the probabilities of the disaster and the risks involved in the declaration of martial law at length. Let's not rehash them now."

Richard LaForge, the National Science Advisor, rose from his chair. "Mr. President?"

"Yes, Dick."

"I don't believe that we were mistaken. While Captain Purdie is probably right about this first event, it's only the first. After all, the crack we feared is in the Ronne Ice Shelf, not the Larsen Ice Shelf. The big event is yet to happen."

"But how long, Dick. How long?"

"I don't know, Mr. President. It could be hours, days or weeks. We need more time to assess the data."

"You know, Dick, Elizabeth is right about the reaction and the backlash we face. If the people hear about this event and the limited damage that will occur along the coast of Africa, they won't believe the damage that could be caused by the breakup of the Ronne Shelf. People will return to the coast in droves. One risk which we did not envision was of a small event preceding the large event by several weeks."

Jason cleared his throat. "Mr. President."

"Yes, Mr. Graham."

"We won't have very long to wait. The computer projections are coming in. I'll put them on the screen." Jason touched a few keys on his console and then walked up to the screen with a laser pointer. "The preliminary satellite data, together with the various seismograph readings have enabled me to project the size and the path of the wave created by the chunk of ice that calved off the Larsen shelf. Our satellites show that the size of the berg, for lack of a better word, was approximately fifty miles long and twenty miles wide. The initial wave front, shown here, was seventy feet high and was traveling at one hundred fifty-six miles per hour. While Captain Purdie is right, he didn't foresee the secondary results of the wave. The wave will shrink to a mere two to four foot swell in the open ocean, but will continue to travel at over one hundred miles per hour. As the wave approaches landfall, it will build back up in the shallows to a sizeable tidal wave of ten to fifty feet in height, depending upon the topography of the ocean floor in the area of landfall."

"Significant damage will occur at first landfall, here, at Cape Town, and along the coastline of South Africa. Damage and death will be extensive in Capetown unless evacuations were completed. Fortunately, the northern coastline of South Africa and Namibia is sparsely populated and damage and loss of life will be fairly limited. In addition, the wave is not large enough to create a second wave of any significance along the Eastern Seaboard. Tidal swells of one to three feet can be expected, but won't be noticed."

"The problem is not the wave heading for Africa. It's the wave heading south, along the Antarctic Peninsula and into the Ronne Ice Shelf. There is no open ocean for the wave to diminish in size. In fact the wave will increase in size as it is channeled by the Antarctic Peninsula and will blast into the Ronne shelf. That impact should occur in approximately one hour. The vast bulk of the Ronne Shelf will shear along this line, from here to here. The wave will blast into the ice pack, jarring loose twelve thousand square miles of ice. Millions of tons of water will be displaced, creating a massive tidal wave."

"I have programmed the computer to simulate the path, speed and size of the wave on the screen. I suggest that we watch the simulation and discuss our options after the simulation is over."

Jason started the simulation. The lights in the Situation Room dimmed. The simulation showed several waves speed out of the Weddell Sea and into the South Atlantic. The waves spread in both directions and headed towards the coastlines of South America and southern Africa. The first wave inundated the Falkland Islands. Cape Town suffered its fate shortly thereafter. The simulation predicted that the wave would penetrate as far as twenty miles inland in areas where high coastal cliffs didn't absorb the brunt of the assault. The wave progressed up into the middle and north Atlantic. While not a threat in the open ocean, where it was reduced to a several thousand mile wide ten foot high swell, the wave sped up the Atlantic towards the population centers along coastal Europe and the United States. Secondary waves followed the initial wave front as the waves bounced off the coastlines of Africa and South America.

As the first swell approached the coastlines of Europe and the United States, the simulation zoomed in on the United States. The wave tore through the Caribbean, wiping out the numerous small islands that didn't have mountain ranges. As the wave front hit the continental shelf, the swells increased to monstrous heights. The wave passed entirely across the southern half of Florida and then smashed its way up the Atlantic Seaboard, destroying the barrier islands along the Carolinas. It channeled up into the Chesapeake, inundating Washington D.C. and portions of Maryland and Northern Virginia before it retreated back into the Atlantic. Practically the entire State of Delaware was inundated. Then, fortunately, the wave began to lose some of its fury. It tore over Manhattan but failed in its attempt to submerge Long Island. The capes of Massachusetts prevented Boston from suffering the total destruction that befell the cities to its south.

The simulation stopped and Jason stood up. A new stationary map appeared on the screen with the coastal regions bordering the Atlantic being outlined in red and yellow. "Ladies and gentlemen, the simulation you just saw is the computer model's best estimate of the effect of the wave fronts created by the breakup of the Ronne Ice Shelf. The areas shaded in red are areas of total destruction. The areas shown in yellow reflect areas where significant damage will occur."

Admiral Branigan stood up and surveyed the map. "Your computer has projected the total loss of every port city on the Eastern Seaboard south of New York."

"That is correct."

"What about my ships in the Atlantic?"

"As long as they are beyond the continental shelves and in deep water, they should be fine. They can easily ride out the swells as they pass under them. The turbulence from secondary waves bouncing off the coastline and into each other, will create a challenge for your crews, but your ships should be able to ride them out."

"So, we've got a couple of hours before the waves hit the Eastern Seaboard."

"That's right, Mr. President."

"Charlie, what's the status of the evacuation?"

"Not great, Mr. President, especially in Florida. Highways are still jammed in the center of the state. We've managed to relocate approximately eighty percent of the population from south Florida into areas along the west coast of Florida in a line north from Tampa into south Georgia. Most of the refugees have relocated into temporary shelters constructed within old military facilities, schools, convention centers and meeting halls. But one thing that is amazing, Mr. President, is the quality and condition of so many of the old army bases. The reports I've received actually indicate that a number of bases closed down decades ago appear to have had new buildings constructed on them over the past few years."

"Really? That's interesting." While Jason's models had not predicted the present crisis, it did predict that millions of people would have to be relocated from coastal regions. Jason smiled as he thought about the arguments they'd had over spending money refurbishing old bases to house refugees. General Maxwell had fought him tooth and nail, but Jason had prevailed.

"As we predicted, as many as five million people ignored the evacuation order. Even using the National Guard and the Army, we couldn't force everybody to move. The death toll in this country alone may reach well over five million, over one million in South Florida alone. We will never know the true total. Our helicopters flying over evacuated cities are already reporting pitched battles between groups of armed looters. In accordance with your orders, we have not sent any troops in to quell the looters."

"They deserve to die anyway." All of the heads in the room turned towards the other end of the large conference table to where the Vice-President sat. To his right sat a smiling General Maxwell and to their left sat General Patrick McGee, the head of the Black Berets. "Look, Geoff, we may as well spill the beans now." The President's face began to redden, but he didn't interrupt. "I, for one, would not have issued the evacuation order. If Mr. Graham is wrong, your head, my head, all of our heads are on the chopping block. Even with the support of General McGee and his troops, we will all be strung up at the nearest tree." General McGee nodded in agreement.

"We should have initiated a selective evacuation, removed the people we need and eliminated the problem that we'll face with an excess population at home and possible threats abroad. Now, it's too late. Assuming Mr. Graham is correct, and for all our sakes, I hope he is, we are faced with feeding, clothing and housing a massive number of refugees that probably won't survive the next decade anyway."

Jason looked around the room to gauge everyone's reaction. Most looked shocked. Elizabeth Savage recovered first. "What the hell do you mean by that, Joel? Are you fucking out of your mind! How could you even consider not trying to save millions of people?"

The Vice-President responded calmly with a slightly twisted smile on his face. He was enjoying this. "Because many of them will be dead within the decade anyway." The Vice-President paused. The blunt statement rendered the Secretary of State speechless, a first in her career. She turned and looked toward the President. He looked furious but still said nothing.

"Go ahead, Elizabeth. Ask the President or our well-revered Mr. Graham. They'll tell you. It's about time that all of you knew anyway. We need to move forward and rapidly. Isn't that correct, Geoff?" Jim Wilson and the other Secret Service agents standing behind the President inched forward, their eyes alert. They looked uncomfortable for the first time in Jason's memory.

All eyes in the room turned to the President. The President

stood, paced back and forth one time and then gripped the back of his chair. He glared at his Vice-President for a moment and then began to speak. "While I disagree with the Vice-President's timing and his judgment, to some extent, he is correct. Almost ten years ago, when I was a Congressman, Jack Watson and Jason Graham came to my office with catastrophic news. They informed me that they had discovered that the planet was in the early stages of a runaway greenhouse effect and that there was nothing man could do to halt it. Civilization as we knew it would collapse within twenty years. After thoroughly checking their results and consulting with then Vice-President Tom Hammond, we set up a double-black program to evaluate and respond to the crisis, Project Runaway. . . ."

⟡

1 0:30 A.M., EST.
VIRGINIA BEACH, VA.

Arthur Jacobson looked nervously at his watch. As a member of the Project Runaway team, he had learned of the impending disaster several days before the public. He had driven down to his beach house to pick up his wife and two children who were on vacation and to pack up a few things, including his CDs outlining how Jeff Nakamura and Vice-President Friedman had known about the runaway greenhouse effect and had conspired to falsify data years before the Project Runaway team uncovered the problem. Arthur had gone along with Nakamura, he'd had no choice. But he had managed to compile a very incriminating report on Nakamura and the Vice-President. He didn't know what to do with it and he didn't dare store it on his computer at the office, but he didn't want to loose it.

Jack Watson had promised a helicopter would pick them up. He'd confirmed it this morning. He had just about given up and yelled to his wife to load the kids into the car when he heard the telltale thwack of rotor blades. He ran into the house to grab his wife and kids. As he walked down the stairs with several suitcases in hand, the door opened. Arthur couldn't believe it when Jeff Nakamura walked in. Jeff smiled at Arthur's puzzled look.

"Hello Arthur. Ready to go."

"Yes. I'd just about given up hope. But why are you here? Did Jack send you?"

"Actually Arthur, Jack didn't send me." Jeff turned his head towards the open door. "Hamilton did." Hamilton Smith and two armed Black Beret privates entered the house.

Arthur put the suitcases down, but didn't quite know what to make of things. His wife came down the steps with their two children and stopped behind them. He put his hand on top of his ten year old's head.

Hamilton walked towards them. "You see, Arthur, we've been monitoring your activities. We know that you've been busy with that computer of yours." He nodded his head towards the guards. They walked forward.

"What the hell's going on here Hamilton." One guard reached for his wife's arm. Arthur took a swing at the guard. The next thing he knew he was doubled up, the wind knocked out of him by the butt of a guard's rifle. His wife and kids began to scream as they struggled with the guards. Hamilton reached down and grabbed Arthur's arm, jerking him up.

"You have a lovely house Arthur. A beautiful view of the ocean from your deck. It must be lovely to watch the waves lap at the beach." Hamilton dragged Arthur out onto the deck. The guards were tying his wife and kids to the rail. Arthur finally guessed what the bastards were up to. He let out a groan and slipped down towards the deck. As Hamilton bent down to jerk him up again, he slammed his elbow back with all his might, aiming at Hamilton's crotch. He felt his elbow impact with muscle and then bone and he jumped up and lunged for one of the guards. With the surprise, he almost succeeded, but a shot rang out and he felt a burning sensation tear through his knee. Blood poured out.

"I knew you had spunk Arthur. That's why we're here." Hamilton nodded to the guards and they began to tie Arthur to the rail. "Now, unless you tell me exactly what we want to know, you and your family will enjoy this view for roughly" he looked at his watch "two more hours."

Jeff Nakamura ran out onto the deck. He spotted the blood. "What happened?"

"Oh, Arthur's a bit spunkier than I thought he'd be. That's all."

Jeff just shook his head. "Arthur, I'm sorry it came to this."

"Bullshit Jeff. You're a son-of-a-bitch." Arthur spat at Jeff, but missed.

"I'm sorry you feel that way Arthur. I considered you a friend at one time. But you didn't see the light. I gave you a chance but you blew it." turning to Hamilton, "I've wiped out the data on his computer. But it appears he recently downloaded several large files."

Hamilton winced and limped over towards Arthur. "It will be my pleasure to find out where he put the files."

"No Hamilton. That won't be necessary. Will it Arthur. After all, you would like to see your wife and children leave, wouldn't you."

"The disks are in my briefcase. Now, Jeff, for God's sake, please, untie them and take them with you." Hamilton nodded to a guard who retrieved Arthur's briefcase. He looked inside, smiled and nodded to Jeff.

Jeff took in a deep breath. "Let's go." He turned and left the deck with Hamilton Smith and the guards in tow, never looking back at Arthur. Arthur screamed until the rotors drowned him out.

12:05 P.M., EST
MIAMI, FLORIDA.

Raul Santiago lay behind the barricade he had built along the front of his store in south Miami. His arm bled from a spot where a looter's bullet had grazed it. The looters moved on after he shot back. There were too many undefended places to loot. As he tore his shirt to bandage the wound, he looked back over his shoulder into the store. The store held everything that he owned. He wasn't

about to leave it behind. He had worked too hard. He had survived
hurricanes and fought off looters in the riots of 2009. He convinced his wife and children to leave, but he wouldn't leave.

After a minute he noticed a strange sound from outside the
store. It sounded like distant thunder, but there had been no
clouds in the sky all day. Curiosity got the better of him and,
carefully holding his shotgun in one hand, he stepped over his
barricade. As he did, the sky darkened. He looked up, in the
direction of the noise, which had now reached a roar, coming
from the east. He couldn't see very far because a dark wall
blocked off his view. As it swallowed the buildings down the
street, he suddenly realized what he faced. He crossed himself
just before it swallowed him and his store.

12:10 P.M., EST.
CAPE CANAVERAL, FLORIDA.

Technicians struggled to launch the last of the shuttles before the
tidal wave hit. They had managed four launches in as many days,
destroying the previous record for turnaround time, and that
with a volunteer skeletal staff. It wasn't that the technicians were
suicidal or dedicated enough to give up their lives. They weren't.
But they would take the risk that the sphere, constructed of new
composite plastics developed in orbit and erected on the base
several months before, would survive.

Technicians not working directly on the launch had used
every spare moment to dig a massive hole in the ground to bury
most of the sphere and to anchor it to bedrock as best they
could. As the technicians watched the flame trail left by the last
shuttle disappear, they sealed the door to the sphere and waited.
They didn't have long to wait. The sphere's instruments relayed
satellite images of the wave as it approached them. They heard a
deep roar as the wave approached the coastline. They strapped
themselves in and waited for impact. When the wave hit, the
sphere shuddered. Several seats tore from their bolts, throwing

their unlucky occupants around. But the sphere held. After ten minutes of being bounced around, the sphere stopped moving. They had survived.

◆

12:25 P.M., EST.
ST. SIMONS ISLAND, GEORGIA.

Buster Jones pushed his shrimper to full throttle as he pulled away from the dock. All of the other shrimpers had headed out to sea the day before. Unfortunately, the engine had blown on the Valeri several days before the evacuation order. Buster hadn't managed to install the replacement as quickly as he'd hoped. He was about a mile offshore before he noticed that his boat had picked up speed. He looked back. No water remained under the dock that he'd left minutes before.

He had a very bad feeling about the situation, but there was nothing that he could do. The current continued to accelerate, pulling him forward at an ever-increasing rate. He couldn't keep track of his speed, but noticed that the land had already disappeared from his view. In front of him he saw a dark line running from horizon to horizon. He didn't know what to make of it. As he closed on it, he began to guess at its actual size. It appeared to be hundreds of feet tall. He said a silent prayer as he closed on the wave. Momentarily, he thought that his prayer was being answered as the wave lifted the Valeri. He thought that it might lift him all the way over the top until the wave began to break. The white foam was the last thing he ever saw.

◆

12:38 P.M., EST.
NEW YORK CITY.

Joey the Knife Rogatano ignored the evacuation orders. He didn't believe the crap he'd heard about a flood. He felt that the whole thing was a put-up by the Feds to round up undesirables and throw them in prison. With his record, Joey couldn't risk it. He'd been a hit man and thief all his life and was responsible for numerous murders, very few of which had been pinned on him. He was a wanted man. Besides, with New York cleared out, he'd have the opportunity of a lifetime to clean out a bunch of uptown residences. When people returned to New York he had a number of places to lay low.

Joey hid out in a pad in the South Bronx while the National Guard patrolled the streets in an attempt to prevent looting and to clear out the last remaining citizens that had refused to leave. Finally, the National Guard had pulled out last night, leaving the city empty of authority. Joey crept out of his safe house and made his way downtown towards a stately old building overlooking the Hudson River that he had scoped out for weeks. He heard periodic gunfire and avoided it, finally making it to the building.

With little effort, he broke into the lobby. He tried the elevators. They were still running. He stopped on the thirtieth floor, the penthouse. The lock on the door did little to delay him. He knew the location of the wall safe and opened it up. The glitter of diamonds and emeralds met his eyes. He smiled gleefully. He would be set for life. He knew that the couple that owned the apartment had been out of town for several weeks and he had hoped that they hadn't returned to remove their valuables. Before hitting the other apartments in the building, he decided to celebrate. He went to the refrigerator and opened it. It contained several bottles of champagne. He pulled one out, Dom Perignon. It sounded French and in this place, it must be good.

He carried it and the jewels out to the balcony to look out over the river as he celebrated. Today he owned New York. As he pried the cork off of the bottle, he heard a funny noise. The sporadic gunfire to the south suddenly died out as the noise grew to a roar. He looked down at the river. There was no water in it. He looked up. A dark shadow moved towards him, swallowing buildings as it approached. Moments before it hit him he realized that he was looking into the face of death.

✦

Reports flooded the situation room. Satellite pictures showed the extent of the damage. South Florida was a total loss. The tidal wave had expanded the deep gouge taken out of the Palm Beach area by hurricane Marla into a broad gulf. Below Ft. Lauderdale, all that remained was an archipelago. Up and down the coast, reports were no better. Jason's computer models of the devastation had proved accurate, although the actual pictures provided by satellites and aircraft conveyed the message far more intensely than had the computer simulation.

The view of Washington sent chills up the spines of most of the individuals in the room. All low-lying areas had been wiped clean of trees and buildings. Only debris remained. The view from one chopper heading up what used to be the Mall showed the foundations of several of the old monuments covered in mud and wreckage. But the ultimate scene they would remember was of a broken section of the Washington Monument driven up against the steps of the Capitol Building, the great dome laying shattered off to one side.

Other cities up the East Coast had fared no better. New York from the air looked like a series of building blocks that a child had set up and then proceeded to tear down. Miraculously, the Empire State Building remained standing while the buildings around her lay in ruins. Perhaps the buildings to the south had shielded her from the blow. The President finally signaled Charlie to shut off the scene.

"I believe that we've all seen enough. We will be seeing plenty more in the days and weeks to come."

The Vice-President remained seated. Even he appeared pale. "What do you propose we do now, Mr. President?"

"What can we do? Pick up the pieces and go forward. The original Project Runaway plan was to complete construction of the first of the space cities before suspending civil liberties. I believe that we now need to deviate from that plan. As horrendous as this event has been, we must use it to our benefit. I will issue an order setting up a program to employ the displaced people for the construction of new cities for them to live in. We can attempt to build arcologies similar to those in Chaco Mesa and Colorado Springs to house the displaced people. It will provide employment for those that are displaced and give them hope for the future."

The Vice-President clearly didn't agree. "But how will we keep them from trying to move back to the coast and rebuild there? After all, unless you go public with the runaway greenhouse effect, people will want to move back to their old homes and rebuild along the coastline. How can we stop them unless we explain that the rise in sea level and continued increase in storms will make the coastal areas unlivable in the next few years anyway? And how will you stop Congressman Gallager and his cronies in Congress? They won't lie still for the expansion of Presidential authority."

"Let's force the issue. Go on television. Announce that we face greater catastrophes in the future due to the runaway greenhouse effect. Declare that martial law will remain in place indefinitely. Announce that you will remain on top of the situation and will work for the benefit of the people. But indicate that dissent will not be tolerated. General McGee has already positioned the Black Berets to maintain control of all key military facilities. We can take control of the country now. Do it!"

The President glared at his Vice-President. Minutes ticked by. Finally, the President spoke. "I disagree. Democracy may be dead already, but I will not dance on its corpse. I do not have all of the answers. But we can not add to the shock the displaced millions have already suffered by announcing the global warming threat now. We will be busy enough trying to feed and house the displaced millions. I'll go on TV in a few hours and announce a plan to rebuild housing for the victims. In the interim, given the chaos, I'll announce the continuation of martial law, wage and price freezes, and the suspension of stock market trading.

"It will take months, if not years, for the people to bounce back from this shock. We can use that time to slowly leak out that the real problem we face is a few years in the future. I would rather not announce that we have known of this risk for years. Too many people would overreact. They'd blame us. If we announce in a few months that new studies indicate that the catastrophe is only a symptom, however severe, of a far worse problem, we can bury our foreknowledge and address the run-away greenhouse effect as a new problem that faces us all. Our prior actions will not be questioned and we will, hopefully, gain the cooperation of our own people and of the people of the world to face the problem in a rational way.

"Admit it, Joel. Even with the Black Berets and the rest of the military, we still need to keep the people behind us. We cannot afford to antagonize the entire population, yet. Our foothold in space remains too fragile. We need a working industrial nation behind us to keep moving forward. The steps you propose would result in the end of a functioning society in fairly short order. We can't afford that."

The President stopped speaking and looked across the table at his adversaries. After a brief pause, the Vice-President responded. "You win, Geoff. I can see your point. But, rest assured, we are ready to take control of the country by force, if necessary."

CHAPTER 28

Fires continue to spread throughout what is left of the Central American rain forests, although rain forest is no longer an apt description. The longest period of drought in record history, combined with temperatures reaching upwards of one hundred and thirty degrees, have created a tinder box of what once was one of the lushest forests in the world. The people had hoped and prayed for rain and thought that their prayers had been answered when a line of thunderstorms moved over the country from the Pacific.

Unfortunately, the storms moved too rapidly to do any good, creating mudslides in areas denuded of vegetation with the lightning triggering numerous additional fires. Massive firebreaks are being dug around cities in the path of the fire; however, given the size of the conflagration, the cities may have to be evacuated due to extremely high temperatures and smoke. The only hope to stop the progress of the fires appears to be a tropical storm developing in the Caribbean. To the south, the fire is already approaching the Panama Canal. Officials hope that the combined firebreaks created in front of the canal, together with the canal itself will prevent the fire from spreading farther south.

In the meantime, millions of tons of smoke are being added to the atmosphere daily. In light of the recent new concerns raised about the greenhouse effect, scientists around the globe have expressed concern that the fires and the smoke will accelerate the process of global warming.

LAWRENCEVILLE, NEW JERSEY.

The third day in a row of one hundred plus degree temperatures set tempers flaring in the temporary refugee camps for former New Yorkers set up on the campuses of Rider College and The Lawrenceville School. Water and food remained in short supply. Air conditioning was nonexistent. The wealthy and many middle class residents of New York had avoided the relocation camps by moving in with relatives or managing to find their own accommodations before the mass exodus to the refugee camps. Fortunately, most of the women with younger children and older refugees were relocated to refugee centers with air conditioning. Unfortunately, what the teams that had set up the refugee camps had not counted on was that those that remain in the refugee camp were, primarily, the young and disenfranchised members of society. That is what Jamal Wilson had counted on.

Jamal grew up in Harlem. He joined a street gang at the age of nine. He quickly climbed up the ranks of command within the gang due to his innate leadership skills and his ruthlessness. He had the forethought to think about what the evacuation would mean.

Most of the other gang leaders had elected to remain in town and hit the wealthy, unprotected sections of town as the evacuation progressed. They assumed that they would have a free hand cleaning up what the wealthy members of society had left behind in their haste to leave New York. But Jamal knew the other gang leaders. They would fight amongst themselves rather than split the vast amount of wealth available to them. He didn't want a part of that battle.

Besides, if the Feds were right, all of the brothers that remained in New York would die. He didn't believe that the tidal wave was a fiction created by the Feds. Why would the Feds make something like that up? If the tidal wave wasn't real, the Feds would be crucified. It didn't make sense. He believed something would happen. He also knew that most people would take their most valuable possessions with them. Those that didn't have anything to take could easily be recruited into his gang.

If the tidal wave didn't hit, the Feds would be in such deep shit covering their asses over the needless evacuation that he and his gang could take advantage of the lack of attention at the camps. They could plunder the camps and then slip away to a bolthole back in New York before the authorities caught on. If the wave did hit and the damage and destruction was as widespread and devastating as predicted, chaos would reign supreme. He and his well-organized gang could plunder those in the camps as well as any residential areas around the camps at will. Given the disorganization and chaos that would follow such a disaster, he had no doubt that he and his followers could find a bolt hole in which to hide.

He ordered his lieutenants to get the gang together and go along with the evacuation, taking as much in the way of arms as they could. He had arrived with over thirty loyal followers at the camp located between Trenton and Princeton, New Jersey. By the morning of third day after the tidal wave and their fifth day in the camp, they had convinced several hundred more brothers to join them in plundering those within the camps and then breaking out of the camps to find new places to live. The Feds had not placed any armed guards around the camps. Due to the tidal wave, the Army and the National Guard had their hands full and were spread very thin.

Early in the evening of the third day after the tidal wave, Jamal decided to strike. He and his original gang members stuck closely together and maintained control of all of their automatic and heavier weapons. They would let the new recruits form the spearhead of their attack. They took over the administration building located in an old stone classroom building. While the new recruits went directly to the water and food supplies, the core gang members broke into the safe that held thousands of food and water vouchers and into the medical tent to steal whatever drugs that they could lay their hands on.

By the time that task was finished, one of Jamal's lieutenants had located a large bus that would hold the original gang members as well as their loot. As the first recruits made their move, the situation in the camp deteriorated into an all-out riot. Camp officials were beaten, raped and killed and fighting broke out amongst several smaller and less organized groups within the camp. While Jamal and his gang could fight their way out, given their organization and weaponry, he didn't want to take that step. Instead, he found that the PA system the school had was still functioning.

He took a microphone and stood atop a wall in front of an esplanade located in front of one of the old buildings, fired his old AK47 several times in the air to get the attention of those around him and began to speak. "Brothers and sisters, why do we sit here and fight amongst ourselves? Why do we shed our own blood? When we were forced out of our homes we were promised good conditions to live in, food, water, air conditioning. Have we gotten any of those things? NO! And why not? Because the Man don't give a shit. He's busy taking care of the rich. If He cared about us do you think that He would have let this riot go on for this long? No!

"Why aren't the police or the Army here? Because they don't give a shit about us? Maybe. We all know that they would just as soon let us kill ourselves off. But we also know that they would stop us if they could. Then why aren't they here yet? Because they can't get here. They are busy with camps like this all over the country.

"Just a few miles down the road are the houses of millionaires in Princeton, New Jersey. Are they living in this squalor? NO! Do they have enough food, water and air conditioning? Yes! Brothers and sisters, what is theirs is finally ours for the taking. There's no one to stop us. Let's stop killing ourselves and go to the houses of the wealthy and take from them. The police and the Army are not here. There's no one to stop us. We can take them by surprise and take what should be ours. Let's go, now! We can take what we want."

Jamal jumped down from the platform. Cheers followed him off of the esplanade. He signaled one of his lieutenants to bring the bus up around the corner. He got in the bus and signaled it to pull out. As the bus slowly made its way through the mass of people standing in and around the circular driveway that led through the campus of the old school, he leaned his head out the door with the microphone. "Let's go. There are cars and trucks behind these buildings. Follow us."

As the bus got to the highway in front of the old campus, he signaled the driver to move out. He then sat in the front seat next to one of his lieutenants and smiled. They headed down the road towards Princeton for a couple of miles until he signaled the driver to turn off on a side street. They stopped several hundred yards down the road where the bus would not be seen. Jamal and several armed gang members walked back to the main street into Princeton. He watched from cover to see trucks and carloads of camp members heading into Princeton. After a few minutes, he signaled his people back to the bus. When they got on the bus, one of his lieutenants tapped him on the shoulder.

"Jamal, what's the deal? We could have gotten there first and gotten the best plunder."

Jamal stared at his lieutenant for a few seconds. "I'm surprised, Hakeem. I thought you'd have guessed. We've already gotten plunder worth far more than money in the food and water coupons and medical supplies. Besides, there will be nothing to plunder. You forget that the President is from Princeton. If any city is defended, it will be Princeton. Sure, they may break into a few houses on the outskirts of town, but they'll be butchered before they get to town. What they're doing is covering our trail. By the time the Feds figure out what really happened and who was behind it, we'll be long gone."

◆

PRINCETON, NEW JERSEY.

As Jamal and his gang disappeared into Pennsylvania, the first of the raiders hit the outskirts of Princeton. They didn't get very far. The President was there. He'd spent the night at his house after having spent an exhausting day viewing the devastation along the Eastern Seaboard. Before he heard any gunfire, his Secret Service team had informed him of the rioting just a few miles down the road. He heard gunfire as he entered Marine One with his family. As his helicopter took off, he could see the trucks of looters stopping several streets up.

In disgust, he turned to Major Chris Beckett, the liaison officer from the Black Berets. "Major, I want that looting stopped." The Major nodded in acknowledgement and got on the radio. He spoke briefly into the radio and turned to the President. "It's taken care of, sir." The President nodded. He was too numb from the dual shock of his air tour and the assault on his home to inquire as to how it would be handled. He regretted his neglect later.

A squadron of Black Berets in helicopters lifted off from a nearby base and flew directly to Princeton. They opened fire on the rioters without warning from several different directions, killing dozens of them. However, the point of the gunfire was not to kill, but to herd the rioters into a small area. Once the main body of the rioters had been packed into a small enough area, the helicopters released a nerve gas that almost instantly paralyzed the rioters. The helicopters landed and the Black Beret commander radioed in for troop trucks to pick up the rioters and take them back to the now desolate refugee camps.

Per instructions received directly from General McGee himself, the commander of the Black Berets waited for several members of the press to arrive before beginning the military trials.

He read a brief statement to the press: "As you can all see from the destruction at this refugee camp, rioting broke out for no apparent reason due to the instigation of several unknown individuals. The immediate effect was the deaths of several relief workers and the injury and rape of dozens more. These relief workers had all volunteered to help out their fellow citizens in dire need. The thanks they received was murder and rape." The commander then brought several injured members of the refugee camp staff up in front of the cameras to show the public the extent of their injuries.

"In addition, after fighting amongst each other, the rioters turned towards the peaceful town of Princeton, New Jersey, to rape and pillage the town. Before being stopped the rioters killed over half a dozen residents and refugees billeted in the houses of those residents. Many more were wounded, raped or robbed."

"By direct order of the President and pursuant to the authority granted by the Riot Control Act of 2012, we have established military courts to deal with the criminals. In order to set an example for the rest of the riffraff prowling through the refugee camps, waiting for the opportunity to prey on the millions of honest citizens living within the refugee communities or in peaceful and unarmed towns surrounding the refugee centers, the military courts have ordered swift and harsh penalties for the perpetrators of the riots. To that end, I ask all of you today to witness that nothing will forestall the swift hand of justice." He nodded his head towards a lieutenant that led two hundred of the rioters, primarily men with a smattering of women, out to the center of the circle of grass in the middle of the refugee camp. The cameras panned in on the prisoners.

Without hesitation, several Black Berets opened fire on the two hundred unlucky rioters. Several of the rioters screamed out, begging for mercy, while others made a run for freedom. None managed to get far. Dead bodies littered the bloodstained ground. Several reporters retched and fell to their knees as the carnage and the odor of nitrate mixed with blood overwhelmed them.

The commander cleared his throat to get the attention of the stunned cameramen and reporters. "This is the justice that others now thinking about rioting can expect to receive. I have sentenced the remainder of the rioters to forced labor. They will help construct new housing for those millions of honest citizens left homeless by the disaster. Predators will not be tolerated!" With that, and before the reporters could ask any questions, the commander and half of his troops made a swift withdrawal in their helicopters. The remaining troops began loading up the several thousand other refugees turned rioters into waiting transport trucks. While the reporters tried, they couldn't get any of the remaining Black Berets to say a word. They did, however, gain additional footage that would appear across the country of the Black Beret guards casually shooting several of the prisoners that attempted to run away from the trucks.

THE FARM.
BLUE RIDGE MOUNTAINS, VIRGINIA.

Hundreds of miles away, at the Project Runaway offices at the Farm, Jason worked away on the calculations to capture a small asteroid and tow it to the manufacturing centers located around the space cities at Lagrange point 5. A knock at the door interrupted his concentration. Captain Tony Healy and Lt. Colonel Bob Ryan entered Jason's office. Neither of them looked very cheerful. Jason knew this wasn't a social call.

Jason stood up from his chair. Nodding his head at the chairs on the other side of his desk, Jason said, "Gentlemen, please have a seat." Rather than taking a seat, both men stood behind the chairs. Jason sat back down. "What can I do for you?" Jason glanced back and forth between his old friend and Captain Healy. Bob's face remained expressionless. He and Jason hadn't seen very much of each other over the past year. In fact, Bob had become a bit too formal and aloof with him lately. While Jason had meant to corner Bob and ask him what was wrong, he hadn't had the opportunity to do so. Although Bob Ryan outranked the Black Beret officer, it was Captain Healy that spoke.

"Mr. Graham, please turn on your television. There's something happening that I believe you should watch." Although a bit puzzled by the request, Jason complied. He flipped on the television and was startled to see a Black Beret major talking to the cameras in front of a scene of destruction at a refugee camp. He turned up the volume. Jason and the two officers watched in silence as the major made his statement. As the Black Beret guards led the prisoners out to slaughter, Jason stared in disbelief. He glanced at the hardened faces of both of the officers. As the shooting began, he reached for the switch to turn off the television. Captain Healy's hand stopped his in an iron grip.

"I would prefer it if you continued to watch, Mr. Graham." Somewhat startled, Jason sat back in his chair as the captain released the grip on his hand. After the cameras panned away from the scene back to the reporter for a commentary, Captain Healy reached forward and turned the volume down on the set. Jason glared at both men. Neither of them reacted.

"Why was this necessary? There was time for a trial."

Bob Ryan responded. "Jason, you of all people should understand. We need to establish a firm hand. We can't afford to lose control of the situation."

"But this, don't you think they went a bit overboard? Just think of the ramifications. While many people will sympathize with the need for swift action and retaliation, to actually see it is something else. The President is going to have a hard enough time with the public and Congress. Do you really think that you can get away with this?" As if to support Jason's statement, he noticed the face of Congressman George Gallager appear on the television. He reached to turn up the volume, but Captain Healy stopped him.

Staring directly in Jason's eyes, the captain spoke. "Look, Mr. Graham. You do your job and we'll do ours. Just keep in mind what you saw here today." With that the captain turned and walked out of Jason's office. Without any change of expression Bob Ryan followed him out. Jason sat puzzled for several minutes before turning up the volume on the set.

". . . a full investigation of this outrageous incident. The country needs time to heal. It does not need to be ripped apart by the outrageous actions of this President. I intend to convene a special session of Congress in Denver to investigate this illegal action and to discuss the rescheduling of the national elections suspended unilaterally by the President. But, let me repeat, this is still America. No one is above the law: not the President, the Congress, or the military. If the President insists on continuing these outrageous and high-handed maneuvers, I regret that he will force our hands. I will personally seek his impeachment."

Jason turned off the television. He'd seen enough. He looked at his watch and dialed Sam's office number on the newest and largest of the space cities, New Washington. After a few minutes the call went through. He smiled as he saw Sam on the viewer. They hadn't been together for several months, since Sam had moved herself, her father and little Jason up to the space cities. Both of them had decided that the time had come to permanently relocate the family to their future home. Jason had remained to continue to head up the Project, but Sam would move the corporate headquarters up to the space cities.

"Sam, it's good to see you. I've missed you." After a few seconds delay in transmission time, Sam smiled back.

"I'm glad to see you, too." But a look of concern crossed her face. "Why are you calling, dear? Is something wrong?"

Jason decided to delay the news briefly. "Does something have to be wrong for me to call my wife?"

"No. But I know you better than that. The last time we spoke, you were engrossed in the asteroid relocation project. You wouldn't stop in the middle to give me a call."

"Unfortunately, you're right. Have you seen the news broadcasts yet?"

"No, I've been busy working."

"Well, you should turn them on. You'll receive quite a shock when you do."

"I will, but give me a quick summary."

"A riot occurred at the refugee. . ." As Jason was speaking, a message appeared at the bottom of the screen that briefly cut his vocal communication off.

NOTICE: ALL COMMUNICATIONS TO AND FROM THE SPACE CITIES ARE BEING MONITORED. COMMUNICATION OF UNAUTHORIZED INFORMATION WILL BE TERMINATED.

"Honey, what's wrong?" Sam asked.

"Sam, did a message flash across your screen?"

"What message? You started to say something about a riot and then audio cut off."

Jason sat silently for a moment. He wasn't sure of the purpose of the communications restriction, but he decided not to put it to a test for the time being. Between his forced viewing of the shooting of the rioters and the communications blackout, he'd gotten the message. While he had value to those now in charge, he also knew that they regarded him as a threat. He wouldn't give them any excuse to act. "Just watch the news. It will speak for itself."

"But why'd your voice get cut off earlier?"

"I'm not really sure. A message appeared at the bottom of the screen indicating that transmission had been interrupted; probably a little technical problem. Just give little Jason a hug for me. Oh, and, by the way, thank our friend Tom for the advice he gave us. Tell him that I'm sure it will come in handy. I love you, dear." Sam looked both puzzled and concerned but did not say anything. She whispered "I love you," blew him a kiss and disconnected the phone.

CHAPTER
29

November 17, 2016.
Pretoria, South Africa.

While recovering from the disastrous effect of the Antarctic tsunami, the people of South Africa have a new threat to face. Unlike the droughts and extremes in temperature suffered across the rest of Sub-Saharan Africa, South Africa has been blessed with temperatures moderated by the surrounding oceans and plentiful rains. Harvests have increased dramatically and a new age of prosperity, which seemed to have emerged prior to the tsunami, is now threatened with another tidal wave, one that threatens to be far more devastating than that which wrecked Cape Town only four short months ago.

The new threat is the wave of refugees streaming southward from the rapidly expanding desert areas in the north, through the sweltering, inhospitable regions around the Congo Basin, past the new small and fiercely territorial tribal groups controlling the few oases of habitable territory left, towards the promised land of South Africa.

Black and white citizens of South Africa alike are taking up arms together to prevent the tide of refugees from swamping and overwhelming South Africa. Although fewer than one in twenty refugees survive the long trek southward, even those few will total in the tens of millions within the next six months.

International agencies have condemned the closing of the South African borders and the arms buildup underway, but the South African government has repeatedly pointed out to its critics that those same critics in Europe and the United States have done nothing to relieve the suffering of Africa. Why should South Africa die to save that part of Africa already abandoned by the rest of the world?

✦

THE FARM.
BLUE RIDGE MOUNTAINS, VIRGINIA.

Now that the orbit of the small asteroid had stabilized, Jason could again relax. He had spent months working with astronomers and astrophysicists here and on the space cities to develop the software for the pellet guns that had to be attached to the body of the asteroid. Each pellet gun, as they called the small electromagnetic launchers that they had attached to the body of the asteroid, had to launch material off of the asteroid at different speeds to adjust for changes in the orbit of the asteroid. Given the difficulty in determining the mass of the asteroid and the effect of changing its orbit and rotation, the pellet guns required constant adjustments. Jason's software provided for those adjustments.

But now, finally, with the call that Jason had just received from Jamie Hampton, the orbit appeared stabilized and the asteroid would arrive at Lagrange 5 in just over three months. The ability to utilize the raw materials on the asteroid would double the productivity of space factories and enable new space cities and solar power satellites to be constructed much faster. Given the speed of the deterioration of the political and economic climate of the planet, Jason was very pleased.

Unfortunately, in thinking about the decline in the political situation, the dark cloud that had hovered around the periphery of Jason's consciousness reappeared with a vengeance. While preoccupied with the intricacies of preparing and debugging the software and calculating orbits, Jason hadn't had time to reflect on the other elements of the crisis.

Fortunately, before his thoughts became too morbid, the telephone rang. He answered it and was surprised to see a strikingly attractive oriental woman with straight jet-black hair in the viewscreen. Something about her looked familiar, but Jason couldn't place her.

"Jason Graham speaking. Can I help you?"

The woman smiled. "Evidently you don't recognize me, Jason. It's Jennifer Chang. Remember, MIT."

Jason looked into the screen again. Recognition dawned on him. "Jennifer, it is you. I hardly recognize you. You look great."

"Thanks, Jason. You do, too."

"How long has it been? Eleven years?"

"Twelve. By the way congratulations on your marriage and your child. When I heard that you had married Samantha, I was very happy for you, although a bit jealous. She's very lucky."

"Thanks, Jennifer. But how are you? Are you married?"

"Yes. You probably don't remember, but I moved to the West Coast after graduation. I met a Japanese man at the company I was working at. We got married about two years after graduation."

"I wish you'd let me know. I was at Cal Tech at the time. I could have made it to the wedding."

"I would have, except the wedding was in Japan."

"Japan?"

"Yes. I'm in Tokyo right now."

"Tokyo. Well, you clearly didn't track me down and call me from Tokyo to reminisce about old times."

"Only partially. My company is one of Japan's leaders in space sciences. As a result, I've been following your work closely over the past few years. Anyway, I heard that you're going to be in Seattle, at the new arcology, tomorrow. I'd like to get together with you to talk about how our companies might work together."

"I'd love to, Jennifer. But I have a very busy schedule. I'm not sure that I'll have time. I wish that you'd called earlier so that I could fit you in."

"What about meeting over lunch? Unless you've changed over the years, you will be eating lunch. Besides, as much as I'll enjoy seeing you again and discussing old times, I have a business proposition to make that I believe will interest you."

Jason glanced at his schedule on the computer screen. He was scheduled to eat with Walter Fitzhugh, public relations manager for the arcology. It wouldn't hurt to cancel that appointment. After all, he hadn't seen Jennifer in twelve years and she was traveling all the way from Tokyo. Regardless of what her proposal might be, he would enjoy the meeting.

"Okay, I'll have to do some rescheduling, but I will free up an hour or so to enable us to get together for lunch."

"Good. If you still like sushi, there's a great Japanese restaurant in the Hotel Nikatta, located inside the arcology. What time is good for you?"

"12:00."

"Then it's settled. I'll meet you in the lobby of the hotel at 12:00. See you then, Jason."

"Bye." The viewscreen went dead.

✦

NOVEMBER 18, 2016.
BEIJING, CHINA.

Governmental edicts have severely curtailed Western news coverage of the devastation suffered in the southern provinces. However, unconfirmed reports are being received from stragglers entering the capital that hundreds of thousands were killed in the typhoon that struck the southern provinces. The storm was only the last in a series of disasters to hit the region. Plagued by record heat and torrential rains, the rice crops have yielded little and rumors of food shortages run rampant. While shortages have not appeared in Beijing yet, some shortages are apparent in a few of the other cities that remain open to the western press.

Unconfirmed rumors exist that the Chinese government has entered into negotiations with several maverick governors of eastern Siberia with regard to buying crops from the now prosperous and fertile regions immediately to China's north. Russian officials have not commented upon such rumors. Satellite reports show the largest increase in Russian and Chinese troop movements along their Siberian border since the conflicts of the late 1960s.

SEATTLE, WASHINGTON.

Jason got off of the plane a little shaky. A strong storm had rocked his plane as it flew over the Rockies. Before he could catch a cab, he heard a scuffle behind him. He looked around and saw a well-dressed, young man being escorted into the back of an airport security car. As the two security guards flanking the man hustled him into the car, he dropped a bag. A number of pamphlets fell out. The car pulled away.

Seeing no cabs around, curiosity got the better of Jason. He walked over to the one security guard that had remained on the sidewalk and bent over to pick up a few of the pamphlets. Jason and the security guard both stood up with a handful of pamphlets. Jason handed his handful back to the security guard.

"What was that all about, Officer?" A bit taken aback by Jason's question and his help, the guard responded after a moment's hesitation.

"Nothing for you to worry about, just one of those Hoseans trying to sneak into the airport and hand out this dribble." The man looked down at the pamphlets in his hand. "By the way, thanks for the hand."

"My pleasure. But, if you don't mind, I'm from back East. What's a Hosean?" The man looked at Jason as if he had just crawled out from under a rock.

"You know, one of the followers of Hosea Johnson, an end of the world cultist." Jason remembered reading a little about Hosea Johnson and his followers some time ago.

"I think I've heard of them, but that man didn't look like a cultist."

"I know. That's one of the problems. They're not like other cultists. They dress like ordinary travelers so that we won't recognize them. Once they're in the airport, they pass this dribble out to passengers in an attempt to win converts and to keep people from flying."

"I hate to appear ignorant, but I still don't see the problem. What difference does it make what they say?" The guard looked at him strangely.

"You're not one of them. . . no, you don't have the look. But it's been all over the news. Where have you been? Ever since the wave wrecked the East Coast, they've swarmed all over the place. And they get results. Airline traffic is down thirty percent. We're being paid to throw them out of the airport. One of them recently tried to plant a bomb on a plane in San Diego. We arrest ten to fifteen of them a day. It's a royal pain in the ass. You can't imagine the additional paperwork they've created for me."

"You look like a reasonable fellow. Here. Take one of these and read about the crap they preach. If anybody approaches you talking about this crap, call security. We'll get to you quickly." Just then the man's walkie-talkie beeped. "Got to go. Thanks again and have a good trip."

"I will. Thank you." Jason took a cab to the train station. Before boarding the maglev train that would take him to the Seattle Arcology, located fifty miles outside of town, he noticed several well-dressed men and women handing out the pamphlets that he had picked up at the airport. Those Hoseans really were out in force.

Once seated on the train, he read over the pamphlet. Jason whistled to himself quietly. Once you plowed through the rhetoric, whoever this Hosea Johnson was, he'd hit the nail on the head. His "revelations" predicted increased storm activities,

droughts and floods, rising sea level and, as the pamphlet pointed out, the flooding of the eastern seaboard. No wonder he had
developed such a following. The pamphlet emphasized that once
you joined the Troop of the Revelation, all your worldly needs,
clothing, shelter and food would be taken care of until the date
when the Lord would descend from the heavens to rescue his
loyal followers from the Hell that he had created on the Earth.

The scary thing about the pamphlet was not only its accura-
cy but its call to arms. While it did not explicitly condone any
physical violence to prevent the atheists and devils incarnate
from fighting God's will, it did call for action against those that
would fight God's plan. Jason could see that the call to battle in
the pamphlet, and presumably in the Reverend Hosea's message,
would lead to an all-out battle against any attempt to stop or circumvent the demise of man due to the runaway greenhouse
effect; what the Reverend called the renewed cleansing of man.
A chill went up Jason's spine as he imagined the number of additional followers the Troop of the Revelation would add once the
fact that the world really was, to some extent, coming to an end
was publicized.

When Jason arrived at the arcology station, he took the people mover to the Hotel Nikatta. He recognized Jennifer in the
lobby. Her figure, which had been good in college, was even
better now. She wore her jet black hair straight and shoulder
length in a way that accentuated the beauty of her oriental eyes
and face. She wore a diaphanous emerald green dress that
looked as if it had just arrived from a Paris showroom. Jason
began to walk over toward Jennifer when she saw him and rapidly walked over to greet him.

"Jason, you really do look great after all these years." Jason
extended his hand towards her. She took it and leaned forward
to give Jason a kiss on the lips. Somewhat startled and embarrassed, Jason backed away a fraction. Why were women he had-
n't seen for years always greeting him in this fashion? Jennifer
laughed at his reaction. "After all these years you didn't expect
me to accept a simple handshake; did you?"

"I wasn't really sure what to expect, Jennifer. It's been a long time and we are here on business."

"You always were a bit too serious for me, Jason. That's one of the reasons that our relationship didn't work out. But don't worry about me. I am here on business, as is my husband. He's meeting with some of the arcology's engineers as we speak. Rather than stand out here in the lobby, why don't we go into the restaurant? I'm starved."

They proceeded into the Japanese restaurant. After ordering, Jennifer continued with the small talk for several minutes. Despite her earlier assurances, Jason would have sworn that Jennifer was flirting with him. It really wasn't like her. An oriental waiter came over to take their drink order. Jason ordered a Japanese beer and then noticed that while Jennifer ordered, the waiter gave her an almost imperceptible nod that Jennifer seemed to acknowledge.

She turned to face Jason, the happy expression on her face replaced by a serious one. "Jason, it's time we talk business."

"I was wondering when you'd get around to it."

"I had to wait until I was sure that we could talk freely. If you don't know it already, my security team informed me that you are being monitored." Jason raised one eyebrow. "I had my security team neutralize the equipment being used to monitor you. Your chaperons will be led to believe that they have an equipment problem, but that only leaves us a few minutes to cover the things we need to. I would have gone into more detail over the telephone, but your line was almost certainly monitored."

"I don't know where you got your information or why, but I think you're right."

"I know that I am." Jennifer leaned forward. "Let me get into the real reason why I called. As a result of your company's success, my company decided to assign a team to study what Whitlock Industries was doing. Due to my connection with you and with Samantha and my background in computers, my company made me head of that team. Jason, I reviewed and studied your Masters Thesis over five years ago. I must say it was a brilliant piece of work." Jason studied Jennifer's expression. He had known her well enough in college and believed that she was sincere.

"Then,. . ." Jason shrugged.

"Yes. We conducted our own studies. They confirm your hypothesis." Jennifer looked down for a moment. "I had hoped that your work was flawed, as we were led to believe when you were discredited; but, knowing you as I do, I knew that you must have been correct."

"We, my company and certain Japanese industrialists and politicians, have known about the runaway greenhouse effect for several years. Like you, we began to prepare. Unlike you, as a result of more limited resources and time available to us, we couldn't attempt to gain a permanent presence in space. Instead, we have been quietly constructing environmentally sealed buildings, like this arcology, in remote locations in Japan and in sections of the Australian Outback. However, times have changed. Given the current geopolitical climate, we would like to join in your endeavor and gain a presence in space to support our Earth-bound shelters in Japan and Australia."

"As I'm sure you've surmised, ever since the Atlantic tsunami destroyed much of the East Coast, your country has been in turmoil. Democracy is all but dead and a power play is occurring at the highest levels of your government." Jason began to say something, but she signaled him to remain silent.

"Jason, my backers want to obtain the cooperation and support of Whitlock Industries to build our own space city and solar power satellites. I see you shaking your head, but we have much to offer. In addition to the technical expertise and skilled workers we can contribute, we offer Whitlock Industries an offshore base of operations in the event that your government shuts down your ground-based operations. In exchange for a small core of experienced workers to train our crews, temporary accommodations on one of your cities and access to the raw lunar materials, we will give you a fully constructed launch facility in Australia."

She finished and took a sip of her tea. Jason sat in silence for a few moments. What she said made sense. Jason had always felt uncomfortable about keeping the United States' allies in the dark when they had so much in the way of resources to contribute. But, given the risk of a leak, the original Project Runaway team had decided not to spread the news beyond a select few individuals in the United States. Now the cat was out of the bag. She was right about the resources they could offer. And, once the asteroid reached the construction areas in about a month, there would be plenty of raw materials to go around. Besides, Jason could feel the Vice-President's grip tightening. The new base would give Whitlock Industries and the new space cities more freedom to deal with the government of the United States.

Jennifer patiently waited for his answer. "You know, Jennifer, I can't think of a single reason to say no. The only problem is that, if I am as closely monitored as you feel, I won't be able to openly communicate with you or with the space cities. Why didn't you go to them directly?"

"Because I know you, Jason. I knew that you would agree to our proposal. There are those that will object. We need your support."

"A lot of good that will do you. They monitor my communications to Sam and to her father. You will need to work directly with Sam, her father and Tom Hammond. I am not in a position to deliver what you want."

"Yes, you are, Jason. That relates to the last part of my plan. I have set up a code that you can use in your conversations with Samantha to enable you to pass messages back and forth without the risk of discovery. A member of my team already delivered a copy of the code to Samantha." Jennifer withdrew a small booklet from her Chanel pocketbook and handed it to Jason. "Here, read this. You can use the code to communicate directly with Samantha. Destroy it after you're done." Jason scanned it briefly and then put it in his pocket.

"That's fine, but they're sharp. Won't they pick up on the code?"

"No. The beauty of the entire scenario is that we're going to give them exactly what they are looking for: a hold over you. I'm sure that they know how close we were at MIT. My husband and I have intentionally had several rather public arguments back in Japan. Given my business, I travel to the West Coast fairly frequently. We will give the appearance to your people that we're having an affair. After all, you and Samantha have been separated for a long time. What would be more natural than for you to have a liaison with an old friend?"

Jason thought about it for sometime. "I don't know. I don't like the idea of faking an affair."

"But, Jason, it's perfect. You'll give them the leverage over you that they want. I'm fairly confident they took a picture of my kissing you in the lobby. That's why I did it. They're bound to record our activities together. Then, at the appropriate time, they'll show the pictures to you to gain your cooperation. If they think they have something on you, they won't keep you on as tight a leash. Besides, if you're concerned about Samantha's reaction, try out the code. If she confirms that she understands, we can start our little game today by you canceling your flight and spending the night with me. My husband is going back to Japan tonight. I'm scheduled to remain for a couple of days."

Jason thought about his run-ins with the Black Berets and his concern over the motivations of the Vice-President and General Maxwell. As had happened over and over again since his discovery of the runaway greenhouse effect, he knew that he had little choice in the matter. Events had made his decision for him. "Okay, I'm willing. But I will confirm this with Sam first."

"Good. I knew that you'd agree. Your last meeting is at the arcology information center. It ends around 6:15."

"That's right." Jason studied Jennifer's expression. She did not bat an eye. "Your information is good. I'm impressed. But for all our sakes, I hope your people are good enough. I've seen what my people can and are willing to do."

Jennifer gave Jason her mysterious smile that Jason used to

love. "Of course they are. We Japanese still hold several technological advantages over the United States. Don't worry, Jason. But we only have a few minutes left before your people become overly suspicious. Tonight, I'll be shopping across the main level from your meeting. I'll make sure to bump into you when you come out. After a brief exchange, I'll suggest that we have dinner. If you want to back out, say no and go on home. If, however, you decide to proceed with my plan, say yes. We'll meet for dinner and go from there."

"Good. . . ." The waiter returned to the table and nodded his head toward Jennifer again. She quickly changed the topic.

"Anyway, I haven't seen either of them in years. But I've rambled too long and we haven't even ordered. I recommend the Ahi. It's always been my favorite." They ordered and ate lunch without discussing anything further of importance. Jennifer continued to flirt with Jason and he, in character, showed a very reserved interest in Jennifer.

CHAPTER 30

With thousands of refugees of the great tsunami still in temporary shelters throughout the countryside, officials fear that many will die as a result of the harshest winter to face England in the last century. Freezing temperatures and heavy snowfalls have gripped both the British Isles and Scandinavia. While these weather patterns appear inconsistent with the numerous recent reports of global warming, key scientists at the British Ministry of Science confirm that the harsh winter is not inconsistent with global warming.

Reports released this morning indicate that the Gulf Stream has slowed significantly and has shifted away from our coastlines. Scientists fear that this is not a temporary phenomena and that the Gulf Stream may never return to its original course. Even with global warming, the weakening and shifting of the Gulf Stream, combined with the rapid shrinking of the arctic ice cap will result in harsher winters for the British Isles and sections of Continental Europe.

In other news, the political situation in the United States has continued to deteriorate. As a result of the tsunami, President Barnes has continued the suspension of the elections scheduled for last November. In response, the United States House of Representatives meeting in Denver, led by Congressman George Gallager, have introduced a bill of impeachment. Preliminary indications are that, unless the President immediately schedules the elections, the President will be impeached. The BBC will cover live the address scheduled by President Barnes for 8:00 Eastern Time.

◆

McLEAN, VIRGINIA.

The streets of Washington's suburbs remained deserted; their pre-tsunami loads a forgotten memory. Numerous neighborhoods lay deserted even though the wave hadn't touched them. Despite the lack of damage, the owners had elected or found themselves unable to return.

Jason drove through several areas hit by the wave. The devastation still amazed him. He drove down one hill and around a curve out of a quiet and seemingly perfect, untouched neighborhood into an area of complete devastation. Not a single tree or house stood below the level swept by the wave. Had the Army Corps of Engineers not cleared and repaired several of the major arteries through Washington's suburbs, many areas not touched by the wave would have remained unapproachable. The neighborhood surrounding the Mansion, EnviroCon's headquarters, was one such neighborhood.

Jason had driven from his home in Middleburg, Virginia, to the Mansion to have dinner with Jack Watson and watch the President's address after dinner. Jason left early to enable him to drive into Washington, D.C., proper, something he hadn't done since the wave had struck. He drove down highway 50 as far as he could and then followed the detour signs. Evidently, the Corps of Engineers had not reconstructed the bridges over the Potomac. The detour led him to a hill overlooking what used to be the Crystal City office complex.

Jason stopped and got out. There was nobody else around. He walked over to the edge of the hill. Somebody had erected a large sign along the edge of the hill. It depicted the location of the various monuments and major buildings prior to the destruction caused by the tsunami. Without the sign, Jason would have had difficulty locating the Mall and the monuments.

The Potomac no longer looked the way it had before the disaster. It had expanded to swallow the reflecting pools along the Mall. Every bridge was gone and only a temporary pontoon bridge had been erected over the new, wider Potomac. Jason looked down to where the water began. He could see that the water now flowed around several large walls on the southern shore of the river. He checked the map and was shocked to find out that he was looking at the remnants of what, before the construction of the arcologies, had been the largest building in America, the Pentagon. The wave had destroyed it as completely as if a nuclear warhead had detonated over it.

Jason could find no trace of the Lincoln or Jefferson Memorials. Although, he could tell from the location of several cranes and other construction equipment that the Corps was working on digging the foundations out from under the mud. The largest broken fragment of the Washington Monument had been moved from where it had come to rest against the steps of the Capitol Building to the stub that protruded out from the mud at its original location. The Dome of the Capitol Building had also been moved back to sit next to what remained of the Capitol Building.

Of the Smithsonian Institution buildings and the White House, Jason could see little trace. He did not know how long he stood looking over the desolate and unfamiliar landscape, but the ringing of bells startled him out of his introspection. He glanced over to see the tower of the Georgetown Cathedral. It, as well as the National Cathedral had escaped serious damage due to their elevation. He glanced down at his watch. It was time for him to leave and meet up with Jack.

He glanced over his shoulder once, returned to his car and drove off. He regretted his decision to come and see the destruction. He would have preferred to remember Washington the way it used to be. Despite the assurances of the President and the current efforts of the Corps of Engineers, he knew that it would never be returned to anything like its prior glory. The President's speech tonight would see to that.

When he reached the front gate to the Mansion, he rang the bell. Several Black Berets walked out from around the wall to check his identification. They signaled for the gate to open and Jason drove on through. Several new M5 battle tanks with Black Beret insignias deployed in front of the Mansion had crushed the pea gravel of the circular driveway into dust. Jason drove around to the rear to where he used to park to find several parked troop trucks with about a dozen Black Berets mulling around them. The old tennis court had been converted to a landing field for an F26 jump jet and several helicopters, all bearing the insignia of the Black Berets.

Jason parked next to several civilian vehicles, including Jack's Cadillac and walked to the main door of the South Wing. A Black Beret sergeant sat behind the old reception desk and greeted him as he entered. He directed Jason to proceed directly to Jack's office in the main section of the Mansion. Jason walked down the familiar corridors, up the stairs, and to the door of his old office, where this entire nightmare had begun so many years ago. Succumbing to temptation, he tried his old door, but found it locked. He then proceeded on to Jack's office.

When he arrived, the door was opened. He glanced in to see Jack standing by his bar, pouring himself a drink. Jack glanced up. "Jason, glad you're here. Come on in. I'll fix you a drink before dinner." Jack made an exaggerated wave with his arm motioning Jason in. The drink in Jack's hand wasn't his first.

Jack's hand moved toward a bottle of vodka. Evidently Jack had decided that Jason wanted a martini. "Not a martini tonight, Jack, just a rum and coke."

Jack's hand wavered for a minute over a bottle of Baccardi

and then reached for a bottle of Mount Gay Rum. "What the hell? I may as well use it up. Besides, you will for damn sure enjoy it more than me."

Jack barely colored the rum with Diet Coke, but Jason didn't complain. The tsunami had destroyed the Mount Gay distillery on Barbados. Jason hadn't tasted any for months. Both men sat down in a pair of large dark green leather chairs, facing the fire that burned in the fireplace. Jason looked over at Jack who was starring into the fire. He remembered the Jack he had first met years before, a Jack Watson on top of the world and in control of his destiny. Now Jack looked haggard and defeated. What remained of his hair was solidly grey. His aquiline chin had disappeared into swaling jowls and his firm, athletic figure had been replaced by flab, dominated by a large pot belly. Jason shook his head and took another sip of his drink.

Just then, Gene, Charles Lane's old butler, walked into Jack's office. In contrast to Jack, Gene seemed ageless. Jason would have guessed that he had been in his late sixties when they had first met, but now he looked no older. In fact he looked younger than Jack. Gene cleared his throat and Jack looked up.

"Mr. Watson, dinner is ready. As you requested, I set it up in Mr. Lane's conference room."

"Thank you, Gene."

Jason and Jack both stood up and followed Gene down the corridor to the small conference room next to Mr. Lane's old office. The small circular burled wooden table was set with two place settings of fine china, sterling silver, and crystal goblets. A large silver tray and cover sat in the middle of the table and a bottle of '97 Brunelo de Montalcino sat opened next to Jack's seat.

Gene pulled the chair out for Jack and Jack sat down. Jason didn't wait for Gene to assist him and sat down at the same time.

"Shall I serve now, Mr. Watson?" Jack looked over at Jason and then nodded. "Yes, Gene."

The old valet pulled the heavy cover off of the tray to reveal a large roast. It smelled heavenly to Jason. Gene slowly carved several pieces for Jack and Jason and served them several side dishes as well. He then poured their wine and silently left the room.

Jack raised his goblet of wine and tilted it towards Jason. "To the past that we have known and to the future we hope to mold."

With some uncertainty, Jason raised his glass. Jack tilted his towards Jason's until they rang together in the way only fine crystal does. Jason took a sip. It tasted exquisite.

"Jack, I don't quite know what all this is about, but this is some feast. When you invited me over to have dinner and watch the President's speech, I thought that you meant we would have a sandwich or something. I haven't eaten like this for years."

"Given what will occur later tonight, it is my farewell dinner to the past; a past that changed forever the day you first walked into this building." Jack took a sip of his wine. "We've been through a lot together and I wanted to celebrate our friendship and our accomplishments."

Jason looked over at Jack. Something was up, but he wasn't quite sure what. "I appreciate it, Jack. God knows that I may not eat like this again for years. But I do hope that once you and I join Sam on New Washington, I'll be able to host you for a dinner like this. I can't promise a '97 Brunelo, but Sam did indicate that they do have several calves growing on one of the farms. Evidently, the implanting of the cow embryos in rabbits has been successful."

"I hope that you're right, Jason. I hope you're right." Jack took a long draw from his goblet of wine, finishing it, picked up the bottle and refilled it.

After both men had a large plate of Baked Alaska, Jack suggested that they return to his office to watch the speech. When they got there, the chairs they had been sitting in had been turned to enable them to watch the screen that had dropped in front of a bookcase on the other side of the office. Two large snifters of cognac sat on the table between the chairs. Jack motioned Jason to have a seat while he walked over to his desk to withdraw two large cigars from the humidor that sat along one corner of the desk. He handed one to Jason and sat down.

Jack pulled a small clipper out of his pocket and clipped the end off of his cigar. He then offered the clipper to Jason. "They're Cuban. Before we released the news about the tsunami, I bought a case. Now that much of Cuba is underwater, they're probably worth a fortune." Jason clipped his cigar as well and took the gold lighter that Jack proffered. He lit the cigar and blew out a cloud of smoke.

"You always did have style, Jack." Jack smiled. He dimmed the lights and turned on the television. Jason sat back in the chair with his cigar, swirled the cognac around in his snifter, inhaled deeply and took a sip of it before focusing his attention on the television. The anchorman for one of the networks was announcing the preemption of regularly scheduled programming for the President's speech. He speculated that the President would react to the Impeachment Bill working its way through Congress but indicated that the White House, as the President's office was still referred to despite the destruction of the grand old edifice, had not leaked any information about the speech.

Jason watched as President Barnes took the podium. "Ladies and gentlemen, I come before you today a humbled man. You might ask: Why humbled? Some might speculate that it is the actions of the leaders of Congress that have humbled me. I wish that were the case.

"I am not humbled by any action of any man. I am humbled by my impotence in the face of nature. In recent years, we have been subjected to the most violent and destructive acts of nature ever faced by this country. During my first months in office, I took a step to prevent the disaster of another Hurricane Marla unleashing its fury against the Gulf Coast. My action, while risky, succeeded. Unfortunately, despite the power of my office and the technology available to us, I could take no similar action to halt the tsunami that destroyed our East Coast. That event humbled me and humbled all men."

"We are slowly recovering from the effects of the tsunami. Even now, crews of the Army Corps of Engineers are busily reconstructing our Nation's Capital. But several weeks ago I received indisputable confirmation of a problem so severe that it dwarfs the problems created by the tsunami."

"We stand at the end of an era. I have the unenviable task of guiding this great country into a new era, an era that will begin in hardship. Ladies and gentlemen, we have indisputable evidence that our planet is in the midst of a major climatological disaster. Speculation has run rampant in the world press about the greenhouse effect. For over half a century people have speculated that man's tampering with the environment would end in disaster. I stand here before you today impotent and humbled by the knowledge that the worst fears of those that speculated about global warming have come to pass. Rather than burden all of you tonight with the details of the problem we face, I have directed that the National Science Advisor release a detailed report on the runaway greenhouse effect."

"You have experienced monster hurricanes and the effects of the tsunami. Unfortunately, those events are merely the tip of the iceberg. In the years to come, we will face more numerous and fiercer storms and unusual or even unheard of weather patterns which will result in major droughts and famines. Ocean levels will rise rapidly to cover many of the low lying areas already devastated by the tsunami. My scientists inform me that we can do nothing to stop the process and little to slow it. What they do say is that we have, at least, a little time to prepare."

"I will need the cooperation and support of all of you to weather this disaster. We all must work together if we are to survive as a nation. We will need to build new housing and transportation systems that will survive the climatic changes that we face. Together we can survive. But we will each need to make sacrifices to survive. We will all need to contribute everything we have if we are going to survive."

"I know that this will be difficult; but together, we can survive. Our forefathers faced a vast and uncharted wilderness when they first landed on the shores of this great land. Yet from their humble and tenuous beginning, they remained undaunted and strove forth to conquer a continent and create this great Nation."

"Our task is as difficult as that faced by our forefathers several centuries ago. Like our forefathers, we will succeed. To that end, I have issued a number of executive orders to properly utilize the manpower and resources that we have at our disposal and have extended martial law indefinitely. I have also suspended all elections for the duration of the crisis. I know that many voices of dissent exist out there and many have been demanding the return of elections. Our government was built to protect the voices of dissent and the minority. However, this great land can not survive this crisis speaking with many voices. Dissent will result in failure; and that we cannot afford. We, all of us, need to work together to succeed."

"Now I know that all of you will have numerous questions about the details of the problem and I will turn over the podium to my National Science Advisor."

Jack turned down the volume on the screen and turned up the lights. "Well, what do you think?"

"He presented it well and he didn't go too far, but Gallager will be all over him. Congress is not going to take this lying down."

"I know, Jason. But, we've taken care of Congressman Gallager. The Black Berets arrested him and several additional members of Congress. There is little Congress can do."

Jason nodded. He knew that democracy had come to an end in America. He only hoped that President Barnes would succeed in keeping anarchy at bay. "I trust Geoff Barnes and I believe that most Americans do as well. But once this sinks in, he'll become the least popular President in history. You and I both know he's right and that he had to take the steps that he did, but the public won't. I only hope that we're ready."

"Don't worry about that. I've seen the contingency plans. We had better hope that the public does follow the President's orders. Let me tell you, they will be given little choice if they do not. And, Jason, we don't have much choice either."

"We haven't had much of a choice about anything since this thing began, Jack."

"That's not what I meant, Jason. The choices we had then

were governed by the discoveries you made. Our choices now are much more limited."

"What do you mean, Jack? I don't quite follow you." Jason straightened his back in his chair.

"Jason, we've known each other for years. I consider you to be one of my best friends." Jack paused for a minute and took a large swallow from his snifter of cognac.

"I know that, Jack; and I consider you to be one of my best friends as well."

"Jason, let me finish. As friends we have always spoken frankly to each other. I'm going to speak frankly now. I didn't invite you over just to enjoy this fine meal or to discuss the President's speech. As we have discussed numerous times in the past, we knew that when the crisis came, we would have to choose sides. I've chosen mine. It's time that you choose yours."

Jason thought about what Jack was telling him, but he still did not get it. He looked questioningly at Jack.

"Damn it, Jason! For a man to be as smart as you are and yet fail to read between the lines." Jack shook his head. "All right. I'll spell it out for you. As I am sure you have seen — you couldn't have buried your head in your work far enough to hide from it altogether — two major factions exist within the original Project Runaway group. The President heads one faction and currently holds sway due to his position. He believes that we should continue to move slowly away from democracy in an attempt to preserve as much of America as we can."

"The Vice-President and General Maxwell head the other faction. They believe that we need to move swiftly and end the sham of democracy. As you probably recall, they would not have ordered the evacuation of the Eastern Seaboard, but instead would have let nature take its course, eliminating in one fell swoop a large section of the population in order to save the rest. The President won out. Now we are faced with an almost insurmountable refugee problem."

"Jack, if I can believe what I'm hearing, it sounds like you support the Vice-President's position."

"Of course I do. It's the only logical position. And while I myself would not have been able to order the deaths of the millions, I can see the beauty in the Vice-President's solution."

"But, Jack, the steps that the Vice-President has suggested are just the steps that you and I, that all of us here at EnviroCon, talked about endlessly before we approached Congressman, now President Barnes. It's one of the things we all sought to avoid, the abuse of the power granted to us by the knowledge we had. We knew democracy would end temporarily, but our whole goal was to preserve the society we had, not to create a dictatorship!"

Jack slammed his drink down on the table, startling Jason. He had rarely seen Jack lose his temper. "Come on, Jason. Get out of that idealistic world you live in! Years ago we were far enough removed from events to academically discuss preserving our society. But events have overcome us. Jason, you've got to see that the President's position is untenable. He has suspended elections. He has arrested several Congressmen. He has become a dictator, yet he pretends, even to himself, to remain the President of a democracy. He would be thrown out of office in a heartbeat if the people could do it."

"But we both know that Geoff's motivations are good. If there is to be a dictator, Geoff's the best choice we've got. I am convinced that he is and will remain a benevolent dictator and will, if and when the time comes, step down and reinstate democracy."

"Jason, you know as well as I do, that's a crock of shit. Benevolent dictator, my ass. You are letting your idealism get the better of you. Sure, in an ideal world, a benevolent dictatorship is, arguably, the best form of government. History gives us several examples of them. But their failing is their lack of longevity. In order to remain a dictator, you must be ruthless and wield your power with an iron fist. Otherwise, somebody else more ruthless than you will replace you. By definition, a dictator must be ruthless. That, in and of itself makes your statement of a benevolent dictator an oxymoron."

Jason grew angry and his face turned flush. He had never

seen this side of Jack before. "Jack, what's happened to you? Have we fallen this out of touch?"

"It's you who is out of touch, Jason. And it's our friendship that made me bring you here, to give you a warning. I gave you one years ago at Chaco Mesa. This is a second. You won't get another. Jason, you are important, but you are no longer irreplaceable. Watch your step or you'll end up like Arthur Jacobson."

"What do you mean, like Arthur Jacobson?"

"I thought you'd heard. He, his wife and kids were killed at their beach house by the tsunami."

"That's impossible. Arthur knew all about the tsunami. He wouldn't … Jack, you don't mean?"

Jack sighed and breathed in heavy. "He didn't cooperate. I wouldn't, couldn't do it. I couldn't have done it to Natalie those many years ago, but there are those that can and will. Jason, I'm trying to save your life!"

Jason thought, my life, yes, but what about my soul? "What is to prevent me from going to Geoff Barnes tonight and having him put a stop to this nonsense right now, Jack?"

"You could try, Jason; and they might let you. If push came to shove, they could take the President down. If you somehow manage to tell him, they will take him down now, even though now is not the best time. The President still has his uses, otherwise he would already have disappeared. They control the Black Berets and a large portion of the regular military."

"You saw the Black Berets outside. You've got to know that our conversation is being monitored. If I can't get a commitment from you tonight, you may not leave here alive." Jason was stunned into silence. This was one of his best friends sitting across from him discussing not only his death but the death of the world that they had worked so hard to salvage. "Unfortunately, Jason, I know you well enough to know that you would give your own life before you would compromise your principles. Would you sacrifice the lives of your wife and child?"

Jason couldn't believe his ears. "You can't be serious, Jack. They're on New Washington. You couldn't touch them."

"Don't be too sure of that, Jason. Think about it. Do you think the Vice-President will remain down here to watch over a dying world? Besides, even before that happens, they would ruin your marriage." Jack reached down and flipped a button on the television screen. The picture changed from a silent press conference to a picture of Jason and Jennifer Chang kissing in the lobby of several hotels, holding hands and entering a hotel room together. While the picture went dark, Jason could hear the sounds of his and Jennifer's moans. "That Jennifer has a good system, but not good enough. While we don't have any pictures of your actual lovemaking, the audio speaks for itself. You wouldn't want these tapes to get to Sam; would you?"

Jason sat in silence, hoping that he had maintained his poker face. He had at first questioned why Jennifer had insisted on all of the electronic gadgets she'd set up in their hotel rooms and the artificial tapes that she'd run. Jason had thought it funny at the time. Now he didn't think so. He was glad that she had taken the steps she had.

"I can see from your expression that you understand, Jason. I hate to be the one to break all of this to you. But I chose to do it rather than a Black Beret major. Jason, as I said earlier, we have no choice. We know that you're not going to work for us. Just don't work against us. When the time comes for you to help us, you'll do it. You have no other choice!"

Jason stood. "Jack, we've been friends for a long time. But, after this, consider our friendship over." Jason stormed out of the room. He knew he was at risk, but doubted that they would have gone to all this trouble if they intended to kill him. They could have killed him at any time. Instead Jason decided that they were setting the stage for the future. By letting him know that he had no choice before they would need him, they hoped that when the time came, his anger would have subsided and he would accept his fate.

Jason couldn't help but smile to himself as he drove away from the Mansion for the last time. He remembered what Jack had said about this being his second warning. The first warning, at Chaco Mesa, had been about the Vice-President's faction and the fears that he and former President Tom Hammond had of the motives of the Vice-President and General Maxwell. Jack was playing his part very well indeed, if he really was playing a part. He had gotten Jason to believe that he had fallen in entirely with the Vice-President's group. No wonder Jack looked so physically worn down. Jason hoped that whatever Jack was doing would workout for him.

CHAPTER 31

MAY 18, 2017.
QUITO, ECUADOR.

Temperatures soaring well into the century mark, combined with the lower levels of oxygen in the atmosphere and high levels of ozone and other pollutants have joined forces to deadly effect. People are literally dropping in the streets. This reporter has witnessed the deaths of several people on this now deserted street. Were I to remove my oxygen mask, I too would have difficulty breathing. As it stands, my crew can remain out for periods of under one hour before the stifling conditions drive us into the hotel behind me. Thanks to a separate generator, it has not lost power or air conditioning.

Rumors abound that armed groups have struck other hotels in an attempt to gain access to power and air conditioning, but I have yet to confirm that. Government troops have sealed off this section of the city, preventing us from moving through the city to see the effect of these stifling conditions. If weather conditions do not improve, this city of several hundred thousand people could see a death toll well into the tens of thousands. While I doubt that very many citizens of this city high in the

Andes understand global warming, they are getting a devastating example of its effects here today. And despite the temperatures, I am forced to wear long sleeves and this broad brimmed hat due to the potential damage to my skin from even a brief exposure to the ultraviolet rays of the sun at this altitude. Quito has had one of the highest rates of death from melanoma in the world for several years now.

✦

10:08 A.M., EST.
RIYADH, SAUDI ARABIA.

Sheik Omar al Faisel shut off the television in disgust. He tired of watching the troubles of others. The people of Ecuador did not know how well they had things. He looked out of his window overlooking the city of Riyadh and the everlasting desert. Heat, they did not know what they were talking about. Temperatures had remained over one hundred and ten degrees for well over a month now. Were it not for the vast amount of capital spent on the construction of energy-efficient buildings and power systems, Riyadh would be in far worse shape today than Quito.

The Sheik could see that Riyadh, that all of Saudi Arabia, was slowing dying, shriveling up like a grape on a vine. Despite the vast wealth and resources commanded by the Saudi royal family, they could do nothing to change the climate of the globe. The oil that had benefited them for so long had finally become more of a curse than a blessing to them. Had not the burning of the oil sold by them contributed heavily to the runaway greenhouse effect that currently confronted them?

Iran and several of their own fundamentalist Moslem clerics already preached that this was the punishment sent down by Allah and by the Prophet to punish those who had let greed and avarice lead them away from the ways of the Prophet. A spark could easily turn into a fundamentalist Moslem revolution.

The Sheik knew that the Americans would turn down the

request that the Saudi government had made to purchase several of the space cities being built. After all, why should the Americans agree to sell the Saudis a city when they will never be able to construct enough to house their own people? The Saudis could only offer money and oil, but what good was that if the scientists' projections were correct. What would the Sheik's cousin the King do when the offer they had made was rejected? With all their military might and with all their wealth, what could they do? They had the power to overcome their neighbors to the south, the west, and even to the north and the east, but what good would that do them. Most of their neighbors were worse off.

But the Sheik did have a plan. He knew the Americans would not sell them a space city, but they would give them one rather than face the prospects of a holy war against America. The mullahs cried out for blood. Let it be the Americans rather than his own people. His agents had already disseminated claims that global warming was a fraud perpetrated by the West to destroy the Arab world. And they were being listened to. How much easier it was to blame America than their own people. When the news reached the public of the denial of the sale of a space city to Saudi Arabia and of the request that Saudi Arabia be granted space aboard the cities for its refugees, the public outcry would be deafening.

That would be his time. Let his cousin sit. He had the support of the Army. He would act. The telephone in his office rang. It was his cousin, the King. The Americans had rejected their last plea and the King had called a meeting of all of his advisors for late that afternoon. The Sheik acknowledged that he would be there. When he hung up, he had a large smile across his face. By late that afternoon, events would already have progressed beyond the control of his poor, weak cousin. He picked up another line, spoke several words into it and then left his office for his command bunker in the desert.

12:12 P.M., EST.
THE CENTRAL MEDITERRANEAN.

Lieutenant Commander Jim Roberts had the Comm. on board the bridge of the Enterprise. A seaman opened the door from the fanbridge and a blast of heat struck the Commander. The door closed quickly, but the heat lingered. It would be several minutes before the air conditioning system compensated for the blast of hot air. He picked up his binoculars to watch the launch of several F-27s for the "exercise" in the Gulf of Sidrah off the Libyan coast.

Given tensions in the Arab world, the Enterprise and the Mediterranean Fleet were on heightened alert. Intelligence reports indicated that the regime in Libya faced an overthrowal and might strike at America in an attempt to avoid an internal conflict. The exercise had been rescheduled from that morning as a result of the meteorologists' prediction of macrobursts in the Gulf. He had been correct; but evidently the Libyans did not have the ability to predict macrobursts. They lost four Mig 29s to that macro burst.

The Commander could understand how the increasing frequency of macrobursts had greatly curtailed civilian air travel over the past six months. He had the good fortune of having been in the Command center when one struck the ship a month before. He remembered the shock as the entire ship vibrated from the macroburst as it washed over the decks of the ship. It had blown both ready aircraft and over a dozen seamen overboard. Only two of the seamen were recovered alive. And the meteorologist had reported that the macroburst had been a relatively minor one. The downdraft had only been clocked at one hundred and ten miles per hour.

The Commander shivered at the thought as he watched the F-27s rise into the air. A loud alarm interrupted his thoughts. He quickly pulled down his binoculars and looked over at the situation board. "What have we got, Trip?"

Ensign Charles "Trip" Washington studied his boards for a minute before answering. "Sir, satellite reports show several small subsonic aircraft entering the eastern Mediterranean. Heat signatures indicate that they are cruise missiles."

The Commander picked up a phone next to him. "Admiral, to the bridge." He put the phone down. "The Israelis?"

"No, sir, I don't think so. Sir! Satellites now pick up the heat signatures of ballistic missile launches coming from Saudi Arabia. Preliminary indications are that the missiles are headed for us."

"Order the fleet to spread out. What ship will the cruise missiles reach first?"

"They should come in over the Ronald Reagan. She's twenty-five miles to our southeast."

"Good! She's got the best antimissile defenses in the fleet. Order her to intercept the missiles. How long have we got?"

"The attack is well timed. The cruise missiles and the ballistic missiles will reach us within a minute of each other."

The Admiral walked onto the Bridge of the Enterprise. The seaman at the door blew a small ceremonial whistle and announced the Admiral's entry onto the Bridge. All of the officers and men on the Bridge not sitting at a critical station stood up at attention and saluted. The Admiral quickly returned the salutes.

"At ease, gentlemen. What have we got, Jim?"

"It looks as if the Saudis have launched an attack against us, Admiral. They appear to have four cruise missiles and two ballistic missiles incoming." The Commander looked at his watch. "We have approximately four minutes and twenty seconds before the missiles are over the closest ship in the Fleet."

The Admiral frowned briefly as he scanned the telemetry boards. "All of our antimissile systems are up and ready to go, Admiral."

"Good." The Admiral pointed to the ship to the southeast that the cruise missiles would be passing over first. "Is that the Ronald Reagan?"

"Yes sir!" The Admiral smiled. Commissioned only two years before, the Reagan was the most advanced Aegis class cruiser in the Fleet. Her arsenal included the new Avenger antimissile missiles as well as the newest high-energy laser systems adapted from the lasers placed in orbit over a decade previously.

"How long before the Reagan's system engages the targets?"

"About twenty seconds for the ballistic missiles, another minute for the cruise missiles, Admiral."

"Any idea what kind of warhead they've got coming at us?"

"No, sir. But intelligence reports indicate that the Saudis possess a sizeable arsenal of nuclear weapons."

The Admiral sat down in the command chair. The Commander continued to pace behind him. An Ensign brought the admiral a cup of coffee. The Admiral sipped it in silence as they waited to hear the result of the Reagan's attempt to knock out the bandits. The silence lengthened and the tension in the Bridge mounted as the Admiral sat calmly sipping his coffee and watching the telemetry display. The dots representing the ballistic missiles continued to approach the spot representing the Enterprise. Clearly the Saudis were gunning for the carrier.

Finally, after what seemed like far longer than twenty seconds, four more blips appeared on the scope rapidly approaching the blips representing the ballistic missiles. The blips merged and disappeared from the scope. The radio operator received the report. "Sir, the Reagan reports splashing both ballistic missiles." A cheer broke loose from the men assembled on the Bridge. As it did, the Commander noticed debris falling into the Mediterranean about three thousand yards to starboard.

The Admiral quieted the Bridge by clearing his throat. "Gentlemen, this isn't over yet." He pointed to the telemetry board. The Bridge grew silent as the men noted the four new blips on the radar representing the cruise missiles. Six more missiles were launched by the Reagan to intercept the cruise missiles. The men watched in silence as two of the four cruise missiles dropped off the display board. But this time, there was no cheer. Two more remained incoming.

The radio operator broke the silence. "Sir, the Reagan reports two bogeys down. The other two bogeys are now in range of the lasers." The Admiral nodded his head and rubbed the coffee cup between both his hands.

Over thirty-five miles away from the Enterprise, the two remaining cruise missiles sped through the air towards their targets at just under the speed of sound. All of a sudden, one of them burst into flame, exploded and fell harmlessly into the Mediterranean. Her fuel tank had blown, but her deadly warhead had not detonated. Seconds later, the final cruise missile began to smoke. However, rather than blowing up, as the other missile had, the lasers burned right through one of the small, stubby tail fins of the missile. It careened off course, almost plunging into the sea before its internal guidance computer compensated for the loss of the tail fin.

It righted itself and headed towards the nearest ship, the Lance O'Kelly, a thirty year old destroyer. Because both missiles had dropped off radar, the Reagan reported both birds splashed and another cheer arose from the Bridge of the Enterprise. However, as the Admiral began to stand up from his chair, the radio operator shouted over the cheering. "Sir, the Lance O'Kelly reports a missile headed in at just over ten miles out."

The Admiral looked startled, but recovered quickly. "Radio the Reagan and let them know that one bird is still loose."

"Sir, they've radioed us. It's too late for them to launch a missile; the target is over the horizon. They can't hit it with the laser. It's flying too low."

The Admiral looked bleak. "Jim, what does the O'Kelly have to stop the missile?"

"Very little, sir. Her only missile defenses are her old Phalanx gatling guns. We'll know in a few seconds whether it is successful."

Warning bells clanged along the Lance O'Kelly as the missile came toward it. When the missile reached within three thousand yards of the ship, the gatling gun let loose. Thousands of rounds of depleted uranium pellets filled the air in front of the incoming cruise missile. Just as the missile approached the cloud of uranium pellets that would tear it to shreds, the warhead detonated. The maker of the missile had designed it to detonate within two thousand yards of a ship in order to counter the Phalanx defense system.

A brilliant flash filled the skies over the Mediterranean. The men on the Bridge of the Enterprise shielded their eyes as the glass in the bridge darkened automatically and steel shutters slammed down. Over the shouting and the radio broadcast to rig for shock wave, the Commander overheard the Admiral mutter "oh shit!" The warhead, while of a relatively small yield, boiled the water for thousands of yards. While not vaporized by the initial flash of heat, the aluminum on the blast side of the ship burst into flames and the iron and steel began to melt. Less than a microsecond later, the blast wave hit the ship, knocking it on its side like a toy. Fortunately, none of the men inside the ship had long to wait for death. Those not killed instantly by the nuclear flash died quickly as the ship sank beneath the boiling waters.

The nearest other ship, over six miles away, suffered serious damage from the blast, but remained afloat. Due to the wide deployment of the ships, only the Lance O'Kelly was lost. On the other ships, thirty seamen died while numerous others were blinded. Luckily, the four planes in the air at the time of the blast were far enough away to avoid damage. The pilots radioed in to find out the extent of the damage and for orders. As much as the Admiral and others on the bridge of the Enterprise wanted to order the planes to retaliate, the Admiral gave the order for two of the planes to fly a recon over ground zero to determine the fate of the Lance O'Kelly while the other two left the Gulf of Sidrah to take up station several hundred miles in front of the Fleet, in the direction of Saudi Arabia.

◈

I 2:28 P.M., EST.
DENVER, COLORADO.

Vice-President Joel Friedman, General George Maxwell and several other members of the President's Security Council listened to the news of the attack in the situation room in Denver. The President was in the air, en route from a visit to the refugee camps along the Eastern Seaboard. He had requested that the Vice-President make similar trips, but the Vice-President had refused.

Now, sitting around the table in the situation room, the Vice-President asked again: "Have we been able to establish communication with the President on board Air Force One?"

"No sir, we have not." None of those present noted the almost imperceptible smile on the face of General Maxwell or the slight nod that the Vice-President gave to him. The others in the room were unaware that a small and elite detachment of Black Berets, under direct command of General Maxwell, had jammed all signals to and from Air Force One. The Vice-President had seen fit to make sure the President could not be reached in the event of a crisis such as the one they now faced. It would be over an hour before the President set down in Denver.

The Vice-President stood up. "Ladies and gentlemen, in light of our inability to contact the President and the gravity of the crisis we now face, as acting President until we regain communications with the President, I am ordering an all-out strike, with conventional and nuclear weaponry, against Saudi Arabia."

Secretary of State Elizabeth Savage jumped out of her chair. "You can't do that, Joel. That's crazy. The world will condemn us. We can't use nukes. We shouldn't even attack until we go through diplomatic channels to find out what's going on."

"Calm down, Elizabeth. You're still living in yesterday's world, not today's. Immediate and dramatic action is called for. We have got to demonstrate to the world that the United States is not going to be bullied or pushed around. You know as well as I do that the Saudis are retaliating for our denying them access to the space cities. Soon our own allies might try similar tactics. We have to convince them not to try it. After all, they may have a chance if they remain on the surface and build arcologies for portions of their populations. But we must let them know that if they try to blackmail us, we will use all the resources at our disposal to destroy them."

"Besides, the Saudis expect us to sit on our butts and do nothing. That's what we would have done in the past. Well,

I'm not willing to sit and wait for the Saudis to apologize publicly while they attempt to build Moslem sentiment against us into a Jihad. We will make an example of the Saudis for the entire world."

The others in the room remained shocked in silence. Even Elizabeth Savage had yet to come up with a response. It was all happening too fast. The Vice-President turned toward General Maxwell. "General, order the Enterprise to launch an all-out attack against the Saudis immediately. I want all of their military installations hit and hit hard. I am personally ordering you to nuke both Riyadh and Mecca. We'll show those bastards who they're dealing with."

Several of the generals and admirals squirmed in their seats. While they might support retaliation, they couldn't condone a nuclear strike against a civilian target. However, most of them had felt the winds of change blow. They would not openly oppose the Vice-President. Elizabeth Savage did not hesitate. "You are crazy, Joel. It's one thing to hit a military target, but to nuke two cities filled with civilians, you must be mad! Especially Mecca. The Moslem world will rise up as one against us. You can't do it."

The Vice-President smiled in reply. "Elizabeth. You still don't understand the new world order, do you? Now sit down and shut up before I have you arrested! Besides, what can the damn Moslems do to us anyway? The seas are becoming difficult to travel. It's dangerous to fly, even with our sophisticated macroburst detection system. They are impotent against us. This way the rest of the world will learn a lesson they will not forget. Now, General, see that my orders are carried out." The Vice-President stood up and left the room.

CHAPTER 32

Temperatures soared into the upper nineties as the combines swept across thousands of square miles of wheat. Only a few years before, the area had been barren, inhospitable tundra. Now, the Russian Secretary of Agriculture controlled the breadbasket of the world. Projections showed that they would harvest a bumper crop. In fact, for the first time ever, the Siberian harvest would boost Russia's grain harvest to the largest in the world. Requests arrived from all over the world to purchase grain from the Russians. Even the United States made a request for grain, having suffered several years in a row of crop failures due to drought.

◆

MOSCOW, RUSSIA.

Nicholai Stipanovich Greschko laughed as he read over the requests and even demands for grain. Publicly, the Russians understated the size of the crop and disputed those western experts that published reports of a bumper crop. Of course, those countries with access to satellite data knew that the Russians had understated their crop sizes, but those countries, other than China, had no real need for the crops. Even the Russian people themselves didn't know the size of the crop. Yes, the Russians had plenty of food for its people, more than enough. Greschko and other leaders of the country knew that the surpluses would not last forever. Soon Siberia would again become a wasteland, but of a different nature than before. Never again would the permafrost cover the lands. Instead, ultraviolet radiation, droughts and heat would turn it into a desert.

Trainload after trainload of food pored into remote and heavily defended environmental bunkers that Greschko and the Russian leadership had set aside for their chosen few. In a relatively few years those leaders remaining on Earth would evacuate to the centers, leaving the population at large to fend for itself. Greschko himself looked forward to a long life on Mars. The base of several thousand people had reached the point where it was, essentially, self-sufficient. Additional tunnels were being bored beneath the surface of Mars even as he sat there, making room for thousands more to emigrate. He would lead and control both them and those few million fortunate enough to get a spot in one of his environmental bunkers, but not lucky enough to be permitted to emigrate to Mars. His chief of security, Sergei Volkov, already in place on Mars, had confirmed that everything was ready for Greschko. His favorite pieces of art had already been removed from museums around Russia for "renovation," and were in transit to Mars to decorate his new residence.

All was ready for his departure. As soon as he solidified his grasp on the environmental bunkers and the launch sites they would support, he would depart. While safety on Mars remained his primary goal, he needed to maintain a power base on Earth were he ever to deal with the American colonies in space. His goals extended beyond Mars. The current crisis left control of the entire human race within his grasp. With the high ground of Mars and strong bases hidden around Russia, he could squeeze the Americans and gain control of the space colonies.

A knock interrupted his train of thought. He scowled but pushed the button, opening the door to his office. Katya Golganov walked in the door and Greschko's original scowl turned into a smile. "Come in, Katya. Come in." Greschko had grown to depend on Katya and almost to love her. Originally, her beauty and the need to drive an additional nail into the coffin of her father had overcome him. His use of her body served as an additional deterrent for his enemies and showed the extent to which he would go. But now Greschko looked forward to taking Katya with him. Long ago he had tired of his sharp-tongued wife and good-for-nothing son. Katya had replaced them in his life. His fondness for her had grown until he had become almost dependent on her.

"Nicholai, General Grishen is on the telephone. He says that it is most urgent that he speak with you immediately." Greschko raised his eyebrow briefly before picking up the telephone. General Grishen was one of Greschko's closest military supporters. Greschko had already transported Grishen's wife and three children to the base on Mars.

Greschko casually turned on the screen of his vidphone. "Yes, Feodor Alexeyevich. What is it?"

"It is serious Nicholai Stipanovich. We received a report that the governor of Irkutsk was assassinated early this morning. A splinter movement of old communists took control of the local radio and television stations to declare the independence of Irkutsk and its secession from Russia. A provisional communist government has broadcast a request for recognition of its new government."

"How the hell did our security people let that happen?" Greschko regretted that Sergei Volkov was not available to handle the situation.

"I don't know. But as soon as I do, heads will roll. You can rest assured of that!"

"What's the present status of the revolt?"

"The nearest base was over one hundred miles away, but

paratroopers are landing as we speak to arrest the 'new government.' They will be dealt with."

"Good. I expect you to take care of this quickly and efficiently."

A major walked up to the general. "Excuse me one moment, Nicholai Stipanovich." General Grishen spent several minutes in a heated conversation with the major before turning back to the phone, his face pale.

"Nicholai Stipanovich, the government of China has recognized its communist brother to the north and has promised to provide military support. Chinese troops are engaging our troops along our borders. Nicholai, you know as well as I do that we cannot stop them short of our use of nuclear weapons! We no longer have the troop strength to do it."

Greschko sat silently for a moment while he considered his response. His face did not disclose any of his thoughts to the general. "General, we are not ready to go nuclear. You have got to hold them for as long as possible. Bloody their little yellow noses and then pull back. We can't afford to force their hand, yet."

"But, Nicholai Stipanovich, we can't let them waltz into the country like this. Next, they'll move to take over all of Siberia. After all, what choice do they have? Their people are starving."

"Don't worry, General. We will deal with those yellow bastards when we are in a position to do it. Implement Borzoi now. Do you understand?"

"Yes, Nicholai Stipanovich. Good luck." The phone went dead. Greschko pushed a button on his desk. A door to a large safe slid open behind the bar in Greschko's office. "Katya, you will find two suitcases in the vault. Bring them out here. After that is done, take this key," Greschko pulled a key out of his pocket and handed it to her. "and open the small safe on the left-hand side of the wall and put the contents of the safe into my briefcase. We will be leaving shortly for Pleseck."

Katya nodded and turned towards the large vault. Greschko dialed several telephone numbers and spoke briefly each time.

By the time he had finished, Katya had returned to the room with the suitcases and his briefcase. He smiled. While she no longer had the innocent glow and striking looks she had radiated when he had first seen her, she remained beautiful. Her youth had departed, to be replaced by a more mature and enigmatic beauty, like that of the rose bud that blooms into full flower, yet remains intriguing due to the variety of hues in its petals. Even the streaks of grey that now peppered her full head of chestnut hair added to her beauty.

Greschko stood up from behind his desk and walked around to take Katya in his arms. As always, she returned his embrace. They kissed briefly. He walked Katya over to the large window behind his desk that overlooked the Kremlin. "Well, Katya, it's time we depart. We leave Red Square behind for a new one that I have built for us, one where I will reign supreme, with powers beyond those that the original builders of the Kremlin dreamed of." Greschko could feel Katya tense up in his arms as they looked out over Red Square. "Do not worry, dear Katya. You will remain at my side. It is you that are going with me, not that cow of a wife of mine. We shall remain together."

Katya relaxed somewhat. She turned towards Greschko. "But what about my things? Do we have time to go by my apartment?" Greschko laughed. Just like a woman he thought to himself. "Do not worry, my little rosebud. I've taken care of you. All that you need has already been sent to Mars. You'll see."

"But. . ." Greschko pressed his finger against her lips to halt her question. "Shhhh, dear. I do not want people to know that we are leaving. It might create a panic in some circles. We cannot go by your apartment." Greschko walked over to his desk and pulled a velvet case out of his top desk drawer. He handed it to Katya. She held it out in front of her staring at it. "Go ahead. Open it. It won't bite. Call it a little going away present."

Katya opened the case and gasped. The sparkle of diamonds and rubies met her gaze. Greschko pulled the necklace out of its case and draped it around her neck, pausing briefly to kiss the nape of her neck gently. "It belonged to the Empress Catherine the Great. I saw it in the Kremlin vaults several months ago and got it for you. Well???"

"Niki, What can I say." Tears began to fall down her face.

"Say nothing at all. It's time to go." Greschko pressed a button on his desk and several uniformed men entered his office.

"Take these suitcases down to the garage and load them into my limousine." The soldiers nodded their heads, saluted, picked up the suitcases and departed. Greschko picked up his briefcase, extended his arm toward Katya, who was taking one last look out over Red Square, wiping a tear from her eye. She took his arm and they left.

Twenty minutes later they had reached the outskirts of Moscow heading for a small military air base. Greschko had planned his departure long before. Rather than depart publicly, he would leave from the small airfield in a small private jet. The jet would take them to the Pleseck launch site where they would ride a space shuttle up to the Russian space station in low earth orbit. He looked over at Katya. She wasn't herself. She looked withdrawn, drooping in the corner of the limo. "Come over here, Katya." She smiled and came closer to him, leaning her head against his shoulder. He stroked her hair gently and she nuzzled against him. He breathed in the scent of her hair. Combined with the excitement of his impending departure from the globe, Greschko became aroused. He pushed a small button on the side panel of the limo raising the glass partition between the rear of the limo and the front seat where his chauffeur and his bodyguard sat. He pushed another button and a curtain drew across the window. The chauffeur and the guard knew that he was not to be disturbed.

He began stroking Katya's hair more vigorously and gently nudged her head down towards his lap. After all these years, she knew his signal well. She lowered her head into his lap, rubbing the palm of her hand against his crotch. She noted that he was already hard. She unzipped his fly and pulled his engorged penis out of his pants. Gently, she began to lick it until he lowered himself slightly in the seat, closed his eyes and began to moan. She took him in her mouth and sucked on his penis until she could feel him begin to tense up. Then, and without warning, she bit down as hard as she could.

Greschko reveled in the pleasure that Katya imparted. He felt himself building towards a climax and tried to hold back but knew he could not. Suddenly he felt a sharp pain mingle with his pleasure and then usurp it entirely. He looked down and saw a pool of blood in his lap. Katya lifted her head up. Blood trickled out of one corner of her mouth where a wry smile had begun to form. She looked into his eyes. He still didn't get it.

When he caught her eyes, she smiled broadly, opened her mouth and spat something out at his face. "You bastard! You son of a bitch! Did you really think that you could get away with it? You killed my father and drove my mother insane. I am not the innocent little girl that you thought I was. You've been waiting for this moment, when you could abandon Russia and take off for your new world. Well, I have looked forward to this moment for longer than you. I have dreamed of this moment ever since I learned of the method of my father's death and every time I lay under your stinking, filthy body. I have planned this for you in your moment of triumph for years."

"Now you, you stinking pig, rather than escape, you shall die like a pig!" Katya withdrew a long sharp knife from behind her and drew it quickly across Greschko's throat. Still in shock, Greschko placed his hand against his throat, attempting to stop the blood pouring from the gaping wound. He attempted to push the emergency button at his side to let the bodyguard know something was wrong, but the unusually strong grip of Katya on his wrist prevented him. He tried screaming, but nothing came out. Slowly, the life ebbed out from within him. The last thing he saw was a small bit of flesh resting on his knee. Just before he died, he realized what he was looking at.

CHAPTER 33

APRIL 14, 2018.
CNN HEADLINE NEWS, ATLANTA.

Hundreds of millions lay dead or dying over vast stretches of Africa. The battle raging on the borders of South Africa continue, but the vast and disorganized hoards of marauders trying to gain entry into the only region of Africa south of the rapidly expanding Sahara capable of supporting itself are being turned away brutally by the armed forces of South Africa. Conditions over the past few years in southern Africa finally resulted in what years of politics could not achieve, the unification of South Africa's blacks and whites. Together they are defending their lands with the ferocity of a trapped bear. And trapped they are. All over Africa, the remnants of armies have poured south to escape the relentless heat and drought that has doubled the size of the Sahara over the past five years.

Cries for help throughout the world have gone unheard as the other nations of the world have tightened their belts for what scientists predict will be a very lean future. However, the President of the United States today is making a last gesture of magnanimity, receiving a shipload of refugees from the desolation of Africa at the newly opened port of Charleston, South Carolina. An old oil tanker filled to the brim with refugees braved the vagaries of the hurricanes running rampant across the Atlantic year-round, was turned away from ports in Europe and, initially, America. But, finally, the President agreed to accept the refugees.

◆

CHARLESTON, SOUTH CAROLINA.

Deep within the belly of the tanker Aquaba, two thin and gaunt looking Arabs huddled in the corner, holding a tattered backpack. Early on in the voyage, one of the gangs aboard had attempted to take the backpack from the Arabs, but without success. Despite their appearances and despite the fact that the older of the two was clearly sick, they fought with training and discipline, killing two members of the gang and disabling several others. After that incident, they were not disturbed on the voyage.

The elder one, hair falling out, would hardly be recognized by a member of his own family, let alone a security team. He had lost over forty-five pounds and most of his hair and teeth as a result of radiation sickness. Only his eyes, as sharp and piercing as ever, could lead anyone to recognize the former head of the Saudi Defense Forces. Sheik Omar Al Faisel leaned over to his companion, "Well, Hakeem, thanks to the will of Allah, we have survived to have our revenge on the American devils." He patted the tattered backpack lovingly. "Soon it will be over and the infidel that ordered the attack on Mecca will die."

"Yes, Your Highness. We survived the attack on Saudi Arabia, by the grace of Allah, and by the grace of Allah, we will do his work." The mere fact, alone, that the ship had not been beset by the endless storms that now raged through the Atlantic, was evidence that Allah was on their side.

"My friend, I am grateful for you having pulled me out from under the fallen pillar in the bunker. But I regret that we will not live to see the results of our work. After all we have been through, it would be nice. But we cannot ask for too much." Both settled back to wait for the bump of the ship against the dock. They would wait, if possible, until the ship had docked and the President had reached his reviewing stand before taking any action. During the tortuous months spent in an Israeli hospital recovering from the broken bones the Sheik suffered when chunks of concrete landed on his legs as a result of the atomic blast that destroyed the air base above his bunker, he and his adjutant and bodyguard, Major Hakeem Deserai had plotted and planned their revenge. They both reasoned that the will of Allah alone had saved them. For if the blast had landed any closer, they would not have survived at all. As it stood, if the pillar of concrete had landed just a foot to the right, it would have killed the Sheik altogether, rather than merely break his legs.

As with most members of the Royal Family, the Sheik had hidden a vast fortune in financial institutions across the world. But unlike many of his relatives, he had also developed several alternate identities that he and the Major used to flee Saudi Arabia after the brief and disastrous war that had destroyed his country. Even in light of the current ecological crisis facing the world, the Sheik found that enough money could still buy just about anything. Through his contacts within the Mosad and his money, he bought himself passage on this refugee ship with the knowledge, gained at an extremely high price, that the ship would not be turned away from America.

Both men felt the ship bang against the pier built where the Charleston Naval Shipyard had stood before the tsunami. Even through the hull of the old tanker they could hear a band strike up a tune that the Sheik knew to be one of the President's favorites, Fanfare for a Common Man. His eyes blazed as the crowd of refugees began to stream towards the stairs that would lead them up onto the deck. The major picked up the tattered backpack with a groan. He then followed the Sheik who walked towards the stairs with the slight limp from the bones that had not healed fully.

✦

The President, glad to be at least making a gesture of charity in the world that he could see falling in around him, smiled at the cameras. He looked at the three piers and the tattering of buildings that were now called Charleston and frowned. What had once been a beautiful city was now no more than a hodgepodge of temporary buildings, an airfield, several piers and the half dozen large geodesic domes anchored to bedrock and designed to survive even the strongest storms. Given the frequency with which massive storms hit the coastal regions, it made no sense to build anything that could not be rapidly rebuilt.

As he looked at the rag tag group of refugees that had begun to disembark from the tanker, the cameras focused in on him. He knew that receiving the refugees was a token gesture only, but it did give the American people an inkling of their former status as the benevolent leader of the world. What neither he nor the cameras noticed was the disappearance of Major Christian Beckett and most of the President's contingent of Black Berets. Over half of the President's guard had withdrawn earlier that morning to a position several miles inland from the coast, seemingly to provide a buffer between the President and the refugee camps several miles inland. They had received reports of a potential riot being led by Hoseans. As the ship neared the dock, the major and the other Black Berets withdrew to their helicopters. As they became airborne, the major spoke into the radio: "The package has arrived." The helicopters withdrew rapidly.

Jim Wilson scanned the crowd with a trained eye. He had strongly advised the President against making this appearance. It would be too easy for the President to be attacked out here in the open. But, of course, the President prevailed in the argument. He explained to Jim the need to make this appearance. Even though he was a virtual dictator, he refused to act like one. Still the consummate politician, he continually campaigned even though he knew that he would never run again. The wall of Secret Service agents remained firm, yet something looked wrong to Jim Wilson. He had a bad feeling about the situation. He scanned the crowd again, and then again. There were none of the ever-present black uniforms in the crowd. Why? Wilson knew the loyalty of the Black Berets to the Vice-President and always kept an eye on them.

He knew that the Black Berets posed as serious a threat to the President as terrorists. Unfortunately, it was not a threat that he could deal with. While he and his agents would fight to the death for the President, he knew that the Black Berets would wipe them out in any firefight. The Black Berets would have to be dealt with by the President. But where were they now? He spoke into the microphone on his lapel. The answer he got disturbed him deeply. He leaned over towards the President.

"Mr. President, there's something very wrong here." President Barnes looked over at him and studied his face. He could read the concern written on Wilson's face.

"What is it, Jim?"

"It's Major Beckett and his gang. They took off in their choppers a few minute ago."

"Took off. . ." The President looked perplexed. Suddenly a look of terror passed over his face. "It's that bastard Friedman. He's up to something. I'd heard reports, but I didn't believe them. Jim, get me out of here, quick." Jim didn't need any additional prompting. He signaled for Marine One to fly up behind the reviewing stand and for the Secret Service contingent to form a tight ball around the President to move him rapidly to the chopper when it landed.

As Marine One moved towards the reviewing stand, Sheik Omar Al Faisel and Major Deserai appeared on deck. They both stumbled as the glare of the sunlight blinded them. As their eyes adjusted and a relief worker assisted them, they could hear the chopper. They looked over at the reviewing stand and could make out the commotion as the President was hustled towards the landing pad. The Sheik knew it was time to act. As planned, Major Deserai handed the backpack to the Sheik, pulled a knife from the leg of his pants and stabbed one of the relief workers. He then broke from the line of refugees and ran towards the gangplank shouting in English, "Death to the infidel destroyers of Mecca," at the top of his lungs. The Sheik waited as the camera swung towards the Major. When a guard at the gangplank attempted to halt the Major, he disarmed him with the training of a martial arts expert, picked up the unlucky guard by the hair, pulled his head back and slit his throat.

The crowd, the press and even the President watched in horror. Again the major shouted his cry of death to the defilers of Mecca. He then raised his head towards the heavens and plunged the knife into his own throat. As the world watched the death of his loyal follower, the Sheik said one brief prayer before triggering the nuclear device in the backpack. He did not live to see his hated enemy engulfed in the fireball that devoured the small hovel that had been called Charleston. Little did he know that rather than kill the destroyer of Mecca, he had unwittingly helped the one that should have been the focus of his enmity.

Thousands of miles away, as the report of the mushroom cloud reached the situation room below the New White House, President Joel Friedman smiled. His plan, conceived of so many months before when his agents had learned of the Sheik's desire to purchase a backpack nuclear device, had succeeded brilliantly. Not only had he gotten rid of the President in a manner that could not be blamed upon him, he had also received the perfect excuse he needed to complete the creation of fortress America.

CHAPTER
34

Hundreds of thousands of Hoseans are traversing the searing deserts of the American southwest towards the thriving Arcologies surrounding the Chaco Mesa Spaceport. Despite temperatures of one hundred and thirty degrees, the Hoseans continue on, their professed goal to convince those living in and around Chaco Mesa to give up their futile attempt to thwart the will of God. Although thousands will die as they cross the desert, they persist in their attempt. The President has ordered that they return to their homes, but the pilgrims have refused to turn back. Troops have been ordered into the area.

The Farm.
Blue Ridge Mountains, Virginia.

Jason sat in his office at the Farm in shock. He had spent last night in his office watching the reports of the death of his friend, President Geoff Barnes. America, already on the verge of economic and sociological collapse, now tottered on the brink. Scenes of violence filled the news reports. The world he and President Barnes had worked so hard to maintain was tumbling down around them. Fortunately, thought Jason, at least the Space Cities were well established and Sam and little Jason were safe. They had accomplished that.

Jason's greatest fear was that, despite their isolation by hundreds of thousands of miles of space from the growing turmoil here at home, the Space Cities would fall into the grasp of the new President. The thought of Friedman and his gang in charge made Jason gag, but there was little he could do. He had already planned his escape during his last 'liaison' with Jennifer Chang. They would meet next at Vandenberg, where her company had a contract with the U.S. military to demonstrate a new bacterial computer chip. The Japanese had genetically engineered what they called the next giant leap in computer technology, a step as large as the move from the early so-called supercomputers to the Omegatron.

While the technology had shown promise, Jennifer told him in confidence that the biological components still had a tendency to deteriorate too quickly. The real reason for the demonstration of the new breakthrough would be to get Jason near the coast where Jennifer and her agents would get him out. They would then head for the new spaceport in Australia and take off for the Colonies before the new President could stop them.

The news broadcasters interrupted their transmission to announce that the President had entered the Press Room at the New White House to hold a press conference. The scene switched to the podium bearing the Seal of the President of the United States in the Press Room of the New White House. Jason watched as President Friedman approached the podium. He looked supremely confident and smug.

"Ladies and gentlemen. It is my sad duty to have had to assume the mantle of the Presidency as a result of the tragic death of my dear friend and predecessor, President Geoff Barnes.

President Barnes will be sorely missed by all of us. He led this country through the most difficult time in its history. But the crisis this nation faces is not a crisis that it faces alone. The entire globe is threatened with disaster. The Third World is dying. Millions have died in Africa and Latin America and millions more will die due to drought, plagues and starvation. Wars rage across all of the continents. The Chinese and the Russians are even now engaged in a bloody conventional war along their border. There is no telling whether that conflict will eventually evolve into a nuclear conflict."

"We have also just learned that the South Africans have given up their attempt to maintain the integrity of their borders. Instead they have detonated several dozen nuclear devices along their borders, creating a deadly nuclear moat between themselves and the teaming hoards that have pressed along their borders. This may shock some of you, but let me make it clear: The United States is willing to take the same or even more drastic steps if necessary to maintain the safety and integrity of our nation. I have, earlier today, ordered the closure of all of our borders. No longer will travel be permitted into or out of this country. People caught attempting to cross our borders will be summarily killed."

"Today I communicated with the current government of Mexico. If they do not keep their people clear from our border, we will be forced to take the same steps as the people of South Africa and create a nuclear moat out of the northern sections of Mexico to keep the tide of refugees from flowing over our borders. We will have a difficult time caring for ourselves. Despite the noble gesture of President Barnes that led to his death, America can no longer support the world. We must take care of ourselves first. And we must make it clear that we will not tolerate threats of any kind. Americans discovered the problems that we face. Americans built the space cities that thrive today above our heads. It is Americans that will salvage this world. But we must do so at home first. Then, and only then, can we return our gaze once more across the two oceans to the rest of the world."

"By executive order I have today extended the terms of martial law across the entire country. I have also ordered a full mobilization of our armed services and the nationalization of all strategic facilities in the country, including the space cities. We are ready to face the challenges in front of us. But we must face them united. . ."

Jason could only stare in disbelief at the screen. He had done it. The bastard had declared his intention clearly. Jason reached for the keyboard to link up with Sam in New Washington when his door burst open. Several Black Berets entered, their guns leveled at him. Captain Tony Healy entered the room behind them.

"I'm afraid, Mr. Graham, that you have to come with us. You are a strategic resource. You are to be relocated to our facility outside of Denver."

"But I was just calling my wife. Can't I let her know what's going on?"

"No. We will let her know."

"For God's sake, my transmissions are monitored anyway! We both know that. You could pull the plug at any time."

"Evidently, Mr. Graham, somebody convinced the President that we cannot take that chance. We've caught onto your little scheme. You will no longer have the opportunity to plot against the President." The Captain threw several photos across Jason's desk. It took him a few seconds to recognize what he saw in the picture as the broken and bloody body of Jennifer Chang. Bile rose in Jason's throat. He vomited into the trashcan next to his desk. The Captain only smiled. "You know, it's amazing what we learned from her. Evidently we grossly underestimated both you and the Japanese. But, believe me, we will not make that mistake again."

The Black Berets escorted Jason out to a waiting helicopter. It flew him to Andrews Air Force Base where he boarded a military transport for Denver.

◆

APRIL 20, 2018.
DENVER, COLORADO.

Jason paced back and forth on the plush dark blue carpeting of his gilded cage. While his silent guards had provided him with food and beverages, he had yet to hear anything from his captors. His inability to do anything frustrated him immeasurably. He ate, slept, exercised and thought. Jason had just stretched out to do a few push-ups when his door slid open unexpectedly.

He looked up, expecting to see one of the four guards that had entered his room over the past few days but was surprised to see Bob Ryan instead. However, Bob no longer wore the blue uniform of the United States Air Force. Instead he wore the black and gold uniform of a general in the Black Berets.

Jason stood up. "Welcome to Denver, Jason." Bob extended his hand towards Jason. Jason did not take it.

"I wish that I could say that I'm glad to be here. What's with the new uniform? Are you going to a funeral?" The smile left Bob Ryan's face.

"Look, Jason, I came in here as your friend. And I'm sorry about the way you were brought here, but it couldn't be avoided. When they found out about your little game with Jennifer Chang, they decided not to take any chances. Besides, you needed to be taught a lesson."

"Bob, she was our classmate, our friend. How could you let them do it!"

A frown briefly darkened Bob Ryan's face, but it quickly disappeared. "It was unfortunate, but you left us with little choice. You've known all along that it would come to this. There is no other way for man to survive."

Jason stared at his one-time friend. He shook his head. "We had hoped. Geoff Barnes and Tom Hammond thought that it could work."

Bob Ryan took a chair. "Face it, Jason. Geoff Barnes was too weak. He refused to use the power at his disposal, so he was pushed aside. As for Tom Hammond, his time will come."

"We are talking about the survival of the fittest, Jason. If you can't cut it, you'll be pushed aside. Joel Friedman has known that from day one. He has bided his time to consolidate his power base and now has stepped forward to grasp the reins of power. And let me tell you, Jason, he isn't about to let go of them. If you or anyone else gets in his way, he'll eliminate you."

"Why am I alive now, Bob? Why hasn't he just had me disappear, like other of his opponents? After all, Captain Healy made it abundantly clear that I am regarded as a threat. Why am I alive?"

A look of concern briefly passed across Bob Ryan's face. He stood up and faced Jason. "Jason, don't make this any harder than it already is. Keep talking like that and you'll end up dead. For now, you are valuable to the President, too valuable to lose. But even you can be replaced if you do not cooperate. Jason, we were friends once and I'd like to think that we still are. Take my advice: Cooperate. You of all people should know which way the wind is blowing. Besides, you have little Jason and Samantha to think about."

Jason grabbed Bob by his uniform and lifted upwards, choking Bob. "What the hell do you mean by that?" Two guards that had remained in the shadows stepped forward rapidly, but Bob signaled them to halt. Jason slowly lowered Bob and released his grip. Bob signaled the guards to leave the room and shut the door behind them. They did so, reluctantly.

"Nothing, Jason. I'm just stating a fact. With or without your help, the President intends to take control of the space cities. After all, he is privy to the same information that we've had for years. He believes in your logic. Rather than sit tight on the surface of the planet and risk retaliation from splinter groups within his power base or rather than face retaliation from other nations, he will rule from the high ground. New Washington is an appropriate name for the city that Sam and little Jason are living in because it is destined to become the capital of the new America, an America ruled by President Joel Friedman. Besides, Jason, things have begun to deteriorate faster than most people think."

"What do you mean? I've been monitoring things fairly closely. My initial projections look pessimistic right now."

"Jason, you're not the only computer genius around.

Remember, you once falsified the NASA databases. What was done once can be done again. You haven't gotten valid information for several years now. Neither has the press. The President has had a team monitor all of the data coming in from satellites as well as from global monitoring centers. Before it's released, the team has modified it. Even the classified data we sent you was altered."

"Ocean levels are rising rapidly, at over one foot a year. Atmospheric levels of oxygen are declining in the equatorial regions. And while the decrease is in tenths of a percent range, the decline is significant enough to make breathing more difficult, especially given the oppressive heat. Projections show that within five years, ninety percent of the population will be unable to function in equatorial regions without the aid of breathing apparatuses; within ten years, that region will spread to cover most of the United States. Ozone depletion has created a serious risk for skin cancer for anybody exposing themselves to the direct rays of the sun outside. Soon even clothing will not provide adequate protection in the polar regions."

"The President isn't going to sit around and wait. Even now, there are barely enough Arcologies built to house the Black Berets, loyal army troops, their families and support staff. People are already dying in droves in the Third World. It will not be very long before that begins to happen here. And when it does, all hell will break loose. Even with the Black Berets and most of the military, we'll still face a serious problem. Just look at those crazy Hoseans; and their ranks have been growing. Why do you think the Chinese are already fighting the Russians? They've seen what's happening in Southeast Asia and in many of their southern provinces. If the leadership of Russia had not started evacuating for Mars, we'd be watching a major nuclear conflict now. As it is, our intelligence sources predict that what has been limited to a major border conflict will deteriorate into a nuclear conflict within three months."

Jason absorbed what Bob had been saying. Little time remained. As predicted by his original models, the wars would only accelerate the process and the decline of man would occur more rapidly. At least the space cities were, essentially, self-sufficient. And to date they were being run in a democratic fashion. While Whitlock Industries technically owned most of the cities and the space factories that had grown up around them, the people were governing themselves. Tom Hammond had moved to and been elected as mayor of New Washington and remained the most important political figure on the cities.

But that would all change if Joel Friedman and his Black Berets gained control of the cities. His best course of action would be to play along with Bob and with Joel Friedman to see what he could do. After all, he believed what Bob said. He could do little good as a corpse.

"I don't necessarily agree with what the President is trying to accomplish, but you are right, Bob. I can see which way the wind is blowing. I don't really have much choice; do I?" Bob nodded his head and patted Jason on the back.

"I'm glad you see things our way, Jason. You always were pragmatic. I'll let the powers that be know that you will work with us and we'll go from there." Bob turned to leave. He then turned back towards Jason, "Oh, by the way, I'll see what I can do for your accommodations. I think that they'll let you out of this gilded cage shortly." Jason only nodded.

After Bob left and the door was closed, Jason sat down to think. He knew that attitudes towards him would not change quite as rapidly as Bob had implied. Clearly after the little deception that he and Jennifer Chang had played on them, they would be on their guard. Jason would have to be extremely careful. He knew that all of his movements would be monitored, but he knew that somewhere along the way he would have an opportunity to act. One last thing that bothered him was Bob. He'd known Bob for years and Bob had not acted quite right. He couldn't put his finger on it, but there was something wrong. Besides, Bob had never called Jason pragmatic before. He had always teased Jason about being overly idealistic.

Jason flipped on the television. The news was on. He saw scenes of several hundred thousand people just on the outskirts of the Chaco Mesa complex. He turned up the volume. The reporter indicated that the crowd had refused to turn back despite the Presidential order and that they were in violation of the decree of martial law. Tanks had moved into position in front of the crowd. The news helicopter zoomed its camera in as the tanks fired several rounds. The shells exploded just in front of the lead ranks of the crowd, spraying them with dust and rock.

The crowd paused momentarily but then continued its move forward. The camera changed to a ground location at the Chaco Mesa spaceport. It showed several jets painted with the Black Beret insignias taking off. The newscaster announced that the crowd would be subdued with gas and then moved to relocation camps. The camera angle returned to the helicopter above the crowd. It showed the jets flying low over the crowds, releasing a yellow colored cloud. The camera zoomed in to show the faces of several members of the crowd as the gas reached them. A man and a woman holding hands began to choke and cough. Their faces turned blue and they began to vomit. The camera remained focused on them. They both fell to the ground and began convulsing violently. The camera then shifted to a wide angle to show the entire crowd as people began to fall to the ground.

What Jason saw horrified him. If that was the effect of harmless gas, he'd hate to see what the President deemed harmful. Jason turned the television off in disgust. Knowing the President's attitude, he had little doubt that the President had just had the entire crowd of several hundred thousand people killed. The man who had argued against the evacuation of the Eastern Seaboard in the face of the tsunami and the man that had ordered the destruction of most of Saudi Arabia, including Mecca, would not hesitate to kill a few hundred thousand religious fanatics. Jason could hear the President's thoughts. They were trouble and would die soon anyway. Why waste resources on relocating them? Jason's resolve hardened even more. He knew that he could not permit his child, hopefully children, if he ever got to see Sam again, and grandchildren to live under the rule of such a tyrant. He would find some way to fight Friedman or he would die trying.

CHAPTER 35

MAY 18, 2018.
LONDON, ENGLAND.
BBC NEWS.

The border dispute between Russia and China in eastern Siberia escalated dramatically this morning when a Russian shuttle carrying diplomats to the Russian space station in geosynchronous orbit failed to reach the station. Radio contact abruptly terminated and debris was located in the orbit where the shuttle should have been. The Russians accused the Chinese of launching the attack from their space station, an accusation that the Chinese denied. However, the Chinese space station suffered a sudden loss of atmosphere when the skin of the station was penetrated mysteriously. Several Chinese astronauts died before the hole was patched. The Russians deny responsibility, but the Chinese blame the high-energy lasers that the Russians erected as a ballistic missile defense over a decade ago.

The Chinese have issued an order fully mobilizing their troops and the Russians are expected to do the same thing. Furthermore, satellite observations indicate that missiles in both China and Russia are in launch position. The minister of defense has issued a statement that the people of Great Britain condemn the hostile acts of the responsible parties and made a plea for cool heads. However, he emphasized that neither Great Britain nor NATO will become involved in the conflict. Civil Defense officials have issued pamphlets indicating the likelihood of fall-out in the event that the Russians and the Chinese resort to nuclear strikes against each other.

On local news, an arrest has been made in the Chunnel bombing incident. Scotland Yard arrested two unemployed Scotsmen for the bombing. They claim their motive was to prevent the infiltration of Great Britain by the starving hoards of refugees roaming throughout Southern Europe.

On a more positive note, the first European space colony was completed today at the L4 point in space between the Earth and the moon. The winners of the lottery for the British emigrants to New London, the name given to the British sector within the space city, have already arrived at the space-port in Covenant. We wish all of you Godspeed.

✦

COLORADO SPRINGS, COLORADO.

For almost a month, Jason had worked on a new program to enhance the capabilities of the United States ballistic missile defense program. Given that he did not see how the program could be used to harm the space cities, he went at the task with new vigor. As preoccupied as he had been with Project Runaway for so many years, he had forgotten what is was like to immerse himself in a new and challenging project. It only took him a few hours to get the old feeling that he used to of melding with the computer. His total emersion enabled him to ignore his captivity in Cheyenne Mountain. He sat back from the computer screen and breathed in deeply, the recycled odor of the air no longer being noticeable to him.

He had worked in isolation on an old Omegatron III that had no link to any other computers and he knew that his activities were monitored. However, as he delved deeper into the programs, he took the opportunity to work on several projects of his own, projects that he hoped he would be able to use if the opportunity arose. He doubted that those monitoring him would be able to pick up the special lines of code he'd hidden in among hundreds of thousands of others. But at last, he was done.

Jason heard the door open behind him. They also knew that he was done. He turned slowly to see which of his captors had entered. To his shock and surprise, he greeted the smiling face of Jeff Nakamura. Jason hadn't seen Jeff in well over five years, ever since Jeff had resigned from Project Runaway and EnviroCon to "retire" from the stress.

Jason stood up and smiled. "Jeff, how the hell are you? It's been a long time."

Jeff strode forward to grasp Jason's outstretched hand. "It's good to see you, Jason. Although the last time I saw you, you didn't have all that grey in your hair."

"Well you didn't have that tire around your middle either." Jeff Nakamura laughed. "But what are you doing here, Jeff? I thought that you'd be off living the good life while it lasted."

"I did for a while, Jason. I moved to the West Coast and took up golf, of all things. I even got married."

"That's great news, Jeff. I'd love to meet your wife." Nakamura's face darkened.

"I'm afraid that's not possible. She died in the '14 earthquake in San Francisco." Both men remained silent for a moment. Jeff had turned his eyes down towards the ground. When he again looked into Jason's eyes, Jason said: "I'm sorry to hear that, Jeff. Real sorry."

"That's old news now, Jason. But I appreciate it. I hear that you have a child of your own." It was Jason's turn to become uncomfortable.

"You're right. But I haven't seen the little guy in several years. You may not know it, but I sent Sam and my son up to New Washington two years ago."

"Jason, that's one of the things I'm here to talk with you

about. You are going to have the opportunity to see Sam and little Jason again real soon." Jeff's eyes studied Jason's. Jason frowned slightly before forcing himself into a poker face. Suspicion had replaced his joy at seeing an old friend again. In seeing Jeff, Jason had forgotten just how hard it had become to tell friends from enemies these days.

"What do you mean, Jeff?"

"You mean you haven't heard?" Jeff looked surprised. Jason couldn't tell whether the surprise was real or not. "The President has decided that the time has come to relocate the White House and its staff to New Washington. Given the situation in Siberia, he decided that he couldn't hold out any longer."

"But how will that involve me?"

"Evidently Tom Hammond, as Mayor of New Washington, and your wife, as President of Whitlock Industries, have refused to accept the President's authority. They've declared their independence from the United States." Jeff Nakamura's face hardened. "Jason, you are going up to negotiate with them on behalf of the President."

"Look, Jeff. I don't know how you are involved in this, but you may not understand my status here."

"I understand it perfectly, Jason. In fact, General Maxwell and I discussed the matter just yesterday. By the way, thank you for finishing the new program for the ballistic missile defense system. We had hoped you would finish before it became time to leave."   Jason was beginning to get the full picture. "I've been monitoring your work, Jason. You haven't lost your touch. You are still the best programmer alive. That's why we'd hate to lose you."

"Look, we know that you are familiar with the security programs used on board both the George Wright and in New Washington. You've got a week to devise a way to disable their security systems and gain access for us to the George Wright and to New Washington."

"And why should I cooperate?"

"Besides your life, how about the life of your wife and child, let alone the lives of the nine hundred thousand or so other residents of New Washington? Jason, you will be on a shuttle to New Washington next week, like it or not. On board with you we'll have a crack troop of Black Berets. We are hopeful that the threat of killing you will enable us to gain access to New Washington. But despite your wife's position on New Washington, we doubt that they'll take that chance. After all, they know the stakes. The President doubts that Tom Hammond will back down. If not, you had better hope that you can find a way to neutralize their security system. If you don't and we don't gain access to New Washington, you'll be there to watch as we blow it out of the sky."

"You can't mean that. You need New Washington. I know that and they know that."

"You're wrong, Jason. While we'd like to keep New Washington, we don't need it. You forget, there are six other large space cities up at L5 now, each of which are self-sufficient. And that doesn't include the European and Japanese Cities at L4 and the older, smaller nonself-sufficient cities. If we blow up New Washington, we're certain that the other cities will agree to our demands. It's up to you to devise a way to neutralize their security systems and gain us access."

Jason thought frantically. "For God's sake, Jeff. You know as well as I do that a week isn't long enough to break that code, even if I had access to it."

"I know that, Jason. While you've been working on the ballistic missile defense program, I've had several of our top programmers working on a copy of the space city's program. We haven't been able to penetrate it. In designing it, you did your typical thorough job. But you know as well as I do that you left a back door into the program. You've always designed them into your programs. It shouldn't take you more than a day or two to determine how best to use your back door to neutralize the security systems. We're giving you a week just for good measure. Oh, and you'll find that copies of the security system programs are now on your computer."

Nakamura turned to leave. "But what if they've changed the programs since I designed them?"

"You'd better hope that they didn't." With that, Nakamura turned and left, closing the door behind him.

Jason sat down in the chair at his desk, his head tilted downward, with his fingers pulling through his hair. Seeing little alternative, he lifted his head up and began to scan through the lines of code on his computer screen.

CHAPTER 36

MAY 23, 2018.
VANDENBERG AIR FORCE BASE, CALIFORNIA.

General George Maxwell slowly stood up from his chair to head out towards the launch pad to view the modifications being made to the shuttle that would take him up to the space cities next week with the President. A specially designed couch had been prepared to relieve the stresses of takeoff for the General. Even the moderate two and a half gs required for the vertical takeoff of the second-generation shuttle were too much for the General's heart that had suffered two heart attacks. The couch would reduce the stress of takeoff to enable him to get into orbit where the care available in the low-gravity sections of the space cities would enable him to live years longer than he would on the planet.

His guards saluted as he left his office for the underground garage of the command center at Vandenberg. There his driver opened the door to his limousine and he stepped into the air-conditioned car. The general ordered his driver to take them up by one of the gates to give him a better view of the old launch sites. As they approached the gate, the general noticed a disturbance. He paid little attention to it. Suddenly, the limousine lurched to one side and then the other, throwing the General violently to the floor of the car. He began to pull himself up, ready to flay his driver when the impact hit. The general felt very little as the other car impacted his and blew up.

Several more cars entered the gate at top speed headed towards the launch site. Before they could get close to the shuttle or the fuel tanks nearby, several helicopters flew over them, tearing into them with 30-millimeter machine guns. Both cars exploded violently, leaving deep pits in the roads where they had been.

✦

COLORADO SPRINGS, COLORADO.

Jason finished his program two days before his deadline. As Jeff had suspected, Jason had built a back door into the program. Knowing that somebody penetrating the system would first have to gain access to the system, Jason had designed the security program to not accept commands through the communications system. Jason had thought about designing the system to not interact with communications at all, but realized that the security system had to be linked to the communications system. His back door was through the communications system. The virus he would plant in the computer to bypass the security system would gain access to the system by being encrypted into the voice messages from his shuttle to New Washington. The system was simple and effective.

Again, he sat back in his chair and stretched his arms over his head. But this time, he didn't smile. This was not a harmless project.

Minutes after finishing, Jason's door opened. As before, it was Jeff Nakamura. Jason figured that the monitors showed his leaning back in the chair, something that he always did when finishing a project and signaled Nakamura that he was done.

"So you've finished, Jason. You don't look as happy to see me this time." Jason didn't answer. "Will it work?"

"Of course it will. You gave me little choice."

"Good. I knew you could do it. That's why I'll be on the shuttle leading the team. You will remain here."

"What do you mean, remain here? You told me that I'd be going up on the shuttle."

"I did, didn't I? But you are far too dangerous and valuable for us to take a chance of losing you. I'll be taking your place."

"But you can't. The program requires my voice to activate it. The penetration virus will be carried along with my voice over the radio. You need me there."

"Unfortunately for you, Jason, that's not true." Nakamura made a signal with his hand. A vague image began to form next to Nakamura. It sharpened into an almost perfect holographic image of Jason. The image spoke: "But you can't, the program requires my voice to activate it." The holograph faded. "Don't you see, Jason? You will be at my side. We have enough recordings of your voice and your mannerisms to have your holograph respond to any questions that might be asked of you. You will remain here while we take New Washington. In a few years, if you cooperate, we may let you come up to see Sam and your son."

Jason's anger couldn't be contained. His face reddened and he ran towards Nakamura, "You son of a bitch!" Before he could get there two burly Black Beret privates had him pinned against a wall. "Relax, Jason. I'll come by before I leave. If you apologize I may take a message up to Sam for you." Nakamura began to laugh when a Black Beret major entered the room at a run. He saluted Nakamura and Nakamura nodded his head.

"General Nakamura, General Maxwell is dead." Nakamura's eyes widened in surprise, but he showed little remorse. Forgetting that Jason was even in the room, Nakamura addressed the major.

"What happened? Tell me quickly."

"A group of Hoseans somehow managed to circumvent our perimeter defenses at Vandenberg and penetrated the gate. Three vehicles, loaded with dynamite, were on a suicide mission to knock out the President's shuttle. One of the drivers must have recognized the general's car and took off after it. Little was left of either the general or his car."

"But what about the shuttle?"

"Our choppers got to both of the other cars before they could get close enough to the launch site to do any damage."

"Damn those Hoseans. Ever since Chaco Mesa, they've been getting smarter and bolder." Nakamura glanced back over at Jason. "Major, let's continue this briefing in my office." Both men departed without an acknowledgement towards Jason. The guards that had been holding Jason released him and left the room as well. Jason rubbed both of his arms to get the blood circulating again. Despite the soreness in his arms, he allowed himself a little smile.

CHAPTER 37

Jason lay in the small bunk in his dark room staring up at the ceiling. His thoughts were far above the ceiling, in New Washington, with Sam and his son. The shuttle with Jeff Nakamura and the Black Berets had taken off hours before. A tear trickled down one side of his face as he pictured them, knowing that he would never see them again. Jason wiped the tear away with his hand as he heard the door open. He sat up. In the darkness all that he could see was a large shape enter the room.

"Jason?" Jason recognized the voice immediately.

"Bob, is that you?" Jason switched on the small lamp next to his cot. It was Bob Ryan dressed in the uniform of a general in the Black Berets.

"Jason, we have very little time, so listen closely." Bob threw a bundle towards Jason. It fell to the floor in front of him.

"Put these clothes on, quickly! I've managed to bypass the security system for a few minutes, but we've got to get out now." Jason slipped on the coveralls that the Black Beret noncoms wore under the Mountain. They had the insignia of a corporal on them. "You'll be my new adjutant. Don't say anything. Just follow two steps behind me. We should be able to get out of here. It's fortunate that you haven't been permitted to move around under the Mountain because it makes it far less likely that anyone will recognize you. The only people that might recognize you are very preoccupied now with the attack on New Washington and the Chinese nuclear strike against Russia."

Jason finished dressing. "Stuff your clothes under the mattress and let's get going."

"But you only disabled the security system temporarily. Won't they notice I'm gone?"

"No, I've used one of their own tricks against them. Look back at your cot." Jason did. He saw himself sleeping peacefully. He smiled as he recognized the hologram for what it was. "Now let's get going."

Jason followed Bob Ryan. They passed through a maze of corridors. Bob walked rapidly, at a pace that indicated to anyone seeing them that the general was in a hurry to get somewhere but not rapidly enough to warrant concern. Bob nodded his head and saluted several times as they walked by several groups of people. Everyone seemed preoccupied. Those personnel not directly involved in the New Washington project must have been watching the blossoming Sino-Russian conflict. There was always the possibility that the conflict could spill over to engulf other portions of the globe.

They entered a section of narrower tunnels and came to a large steel door with a keypad next to it. Bob began to tap several numbers on the keypad and the door swung open with a hiss. Bob stepped into a dimly lit passage carved out of bare rock. Jason followed right behind him. The door sealed itself shut behind them. After walking about a hundred feet, they came to another steel door. This one opened with a round wheel. Bob began to turn it, but it would not budge.

"Give me a hand. We can't go back the way we came and this is the only way out. Pull hard." Jason and Bob both gripped the wheel firmly. For an instant, Jason thought that it was not going to budge. Suddenly the wheel gave way and it was all Jason could do to keep from falling. When they had opened the door, Jason saw that it led to a small chamber that had metal rings buried into the wall leading up. Bob and Jason began to climb. Not being in the shape he used to be in, Jason was breathing heavily when they reached the top. As they climbed over the top rung, a small cave opened up in front of them. Bob led the way to the end of the tunnel and another steel door. This time the wheel turned easily and the door slid back. They stepped outside.

Jason looked up through tall pines at the star-lit sky. He looked in the direction of L5, knowing he couldn't see the space cities, but looking nonetheless. The door closed behind them. "I know that it's a nice view, Jason, but we've got to hurry. There's no telling how long it will be before we're missed. We're about three miles from the main entrance to Cheyenne Mountain, outside the main compound. This is one of the old emergency exits from the Mountain. Not very many people know about it anymore. I discovered it years ago when I was assigned here." Jason nodded. "Now let's go. I hid a motorcycle near here. It should be in this direction."

NEW WASHINGTON, L5.

Sam was in her office watching developments as they unfolded on the planet below. Using their own satellites as well as pirating signals off others, Sam and the citizens of the space cites had a good picture of what was going on down below. Sam and the crisis team had been meeting for weeks as the crisis unfolded. Now she was watching as mushroom clouds sprang up all around the Chinese-Russian boarder. To date, the Chinese and the Russians had limited themselves to using tactical nukes in the theater of conflict. Tom Hammonds' intelligence advisors projected that the war could go total at any time.

A buzzer sounded as Sam's door opened. Tom Hammonds walked in. "What is it Tom?"

"Sam, I've got Jason on the line."

"What? Where? Are you sure?" Sam hadn't heard from Jason in weeks, not since they'd learned of Jennifer Chang's death. Tom had told her that he'd been taken into custody. She'd feared that he was dead.

"Yes, Sam. He's on board a shuttle on final approach to New Washington. He'll be on station outside the docking port in just a few minutes."

Sam smiled broadly and jumped up from her chair. Then she noticed Tom's demeanor. His shoulders slumped and creases traversed his brow. "Tom, that should be great new. What's wrong?"

"He's onboard a U.S. shuttle loaded with troops. As best we can tell, he's a prisoner."

Tears began to well up in Sam's eyes, but she suppressed them. "You know, over the past few weeks, I've imagined all sorts of things. That Jason was dead, that he was being tortured, but my worst fear was that they'd use him against me. God damn those bastards!"

Tom Hammonds took Sam in his arms. "I'm sure he will." Tom pushed Sam away from him. "Sam, we need you. We need to see if Jason's trying to tell us something. We may have to destroy the shuttle. They've easily got the weapons on board to destroy the city."

Sam nodded her head and followed Tom towards the communications center of New Washington.

✦

CHEYENNE MOUNTAIN, COLORADO.

It took Bob five long minutes to find the motorcycle. They jumped on. Bob pushed the machine hard. One thing that he could say for the Black Beret uniforms, they were perfect camouflage on a moonless night.

After having driven for several hours, Bob pulled off the side of the road where there was some heavy cover and stopped.

"Let's rest here for a minute, Jason. We need to talk."

"I agree. We've got a great deal to talk about."

"First things first though, Jason. We still don't have a lot of time. Once they discover us missing, they'll use all of their available resources to track us down. But before we go any farther, I need to know where to go. Did you manage to find a way to stop them from taking over New Washington?"

Jason remained silent for a minute before answering. "Bob, we go back a long way and I'd like to trust you. But you've been on both sides so many times that I don't know whether this escape is for real or not. For all I know, you arranged the whole thing to get me to tell you whether I did anything to screw up the plan. For the time being, I'd rather keep silent."

Bob smiled and then laughed out loud. "After what you've been through, old buddy, I don't blame you for being a bit paranoid. Don't worry. You'll see soon enough that this isn't a trick. I'll show you why I asked. But, for the time being, you are going to need to trust me."

"Do I have any choice?"

Bob laughed again. "I guess not. Let's get going. We can discuss our escape plans when we get to the next stop."

They climbed back on the motorcycle and drove off down the road. It was getting late and Jason noticed he was nodding off. Even the wind didn't keep him from wanting to fall asleep. He tapped Bob on the shoulder and yelled in his ear. "How much farther? I can barely keep my eyes open."

Bob nodded. "Not much. Just hold on. We'll be there in a few minutes." Jason kept his eyes opened by looking ahead. He saw something off in the distance that woke him up. It looked as if they were heading for a military base. Perhaps Bob wasn't on his side. Several minutes later, they were close enough to see that the fence around the base was dilapidated and had actually fallen down in several places. This must be one of the old deserted bases.

Jason was surprised when Bob turned into what used to be a gate at the base. Grass had grown up through the pavement in many places and the ride was rough. They came to a runway.

Bob drove down it until they reached an old rusty hangar. Bob pulled up alongside it and got off the motorcycle. Jason followed suit. Bob signaled for Jason to follow him. He came to a door with a padlock on it. He pulled out a key, opened the lock and opened the door. He signaled Jason to step through.

What Jason saw startled him. The dilapidated old hangar was a ruse. Inside the hangar was brightly lit, with computer equipment on one side and with a streamlined aircraft sitting in the middle, fuel lines running from it. A very attractive woman in an Air Force major's flight suit walked over to Bob, grabbed him in a bear hug and gave him a kiss, which Bob returned with vigor.

Bob turned towards Jason. "Jason, come over here. This is a friend I'd like you to meet. Jason, meet Major Laura Ramsey. Laura, Jason Graham." They exchanged greetings. "Laura was the best wing man I ever had. We've kept in touch over the years and have planned for this night for a long time. Both of us saw what was happening. With my contacts with General Maxwell, I joined up with the Black Berets while Laura remained a flyer. We managed to fake a crash years ago and Laura brought this beauty over from Vandenberg. It's ready to take all of us into orbit. That is, if we have anyplace to go. Jason?"

✦

NEW WASHINGTON, L5.

Samantha Whitlock, Tom Hammonds and the New Washington crisis team were caucusing in a conference room adjoining the communications center.

"We can't take them out. Their shuttle is hardened and they've got lasers to knock out any missiles. If we use our lasers, they can punch a hole in the side of the city big enough to destroy us well before our lasers will penetrate their shielding. We've got fifteen minutes before their deadline. Any suggestions?" Tom Hammonds looked around the room. Sam stood up, her knees shaking. She knew what she had to do.

"Tom, we can't give into them. I know that they'll kill Jason if we don't give in, but we can't. Jason wouldn't want us to."

"Sam, I know how hard this is for you, but your sacrifice, Jason's won't do any good. They'll kill Jason to show their resolve, then they'll try to board the city. If they fail, they'll destroy us."

Vice Mayor, Rhea Reid, jumped up. "But they can't, they need the city. They'll kill Jason. If we don't give in they'll try to breach our hull. We can push them out. But they won't risk destroying the city."

Tom Hammonds strode over to Rhea. "You don't know Joel Friedman like I do. He'd like to get us in tact, but he'll make an example of us to cower the other cities into giving in. There are ways he could kill us all without doing too much damage to the city's structure. He could then repair the city fairly quickly once he takes over one of the other cities. We've got to do something."

Sam was thinking hard. She didn't see a way out without loosing Jason. She looked up. "Tom, I've got a way. We'll let them dock and then blow the docking section. It'll damage the city, but we'll survive."

"But Sam, they're too smart. They'll expect a trick."

"Not if I'm there to meet them."

"Sam, you can't do it!"

"I've got to Tom. Besides, I might get a final embrace from Jason before I go."

ABANDONED AIR FORCE BASE.
NORTHERN COLORADO.

Jason studied the aircraft. He still couldn't believe his eyes. "Bob, I still won't tell you what I've done, but let's get going. You'll have to trust me on this one."

"Okay Jason, I will. Laura, are we ready to go?"

"We will be as soon as you change out of that uniform and into your old blues."

"Then we're set."

"But, Bob, surely, with all that is going on, Cheyenne Mountain will pick us up on radar and shoot us down."

Bob looked over at Laura and they both laughed. "You are paranoid, Jason. And we're not that stupid. This vehicle has active stealth capabilities. They'll never see us."

About twenty minutes later they had opened up the hangar doors and climbed into the aircraft. Laura sat at the controls with Bob in the copilot's seat and Jason at the engineer's seat. They taxied out of the hangar and then accelerated down the old runway. The slowness of their ascent surprised Jason. He spoke into his headset. "Bob, why aren't we going ballistic or at least up at a steeper angle?"

"If we did, our heat signature would give us away. This way we'll take a little longer, but we'll be up in orbit without anybody knowing any better." Jason had no choice but to sit back and enjoy the ride. He actually managed to dose off. He didn't know how long he'd been asleep but woke up with a pain in his hand. It was resting at eye level jabbing against a control clamp with a rough spot. They were in orbit, in zero g. He flipped on his intercom.

"Where are we?"

"Not too far from the George Wright. We'll be there in about half an hour. Now Jason, tell me: What did you do to stop the takeover of New Washington? If you look at your watch, you'll see that the shuttle with Nakamura on board was scheduled to dock thirty minutes ago."

"Well, I used my back door to gain access to the security system. All indications would show that the shuttle was approved for docking. But I also encoded a message that signaled the solar furnace to rotate so that it faced the path of the shuttle as it approached the dock. The shuttle should have heated up so quickly that they wouldn't have realized it until their tanks burst."

"But weren't you concerned that Nakamura would pick up what you'd done?"

"Not really. Most of what I did was trigger codes that I'd

already worked into the security system. Even if he saw the code, what I did was bypass a safety system on the solar furnace. He couldn't have recognized that as a hostile move. Besides, he knew that I thought I would be on the Shuttle. I doubt he believed that I'd kill myself."

"But what about retaliation by the President? How will we stop him?"

"Nakamura and the President realized the importance of the high ground. They've blown their best shot. Even without me I believe that Tom Hammond will be able to protect himself. And with me, well, I placed a back door in the orbital ballistic missile defense programs. As soon as I get to New Washington, I'll be able to take control of them away from the President. Without them, he'll be unable to budge off the ground."

The news of the destruction of the shuttle greeted Jason, Bob and Laura after they docked at the George Wright. They radioed ahead to New Washington to let Sam know that Jason had not been on the shuttle despite the appearance of the holograph. Eighteen hours later, Jason was in Sam's arms at the dock in New Washington. After spending an endless time hugging and kissing, Jason pushed Sam gently away.

"Where's little Jason and your father? I thought they'd be here to meet me."

"I wanted you all to myself, darling. But now let's go meet them. I have a surprise for you." Jason and Sam got on a subway to ride from the shuttle dock to a park at the center of New Washington. When they got out of the subway station, Jason gasped as he looked out over New Washington. Stretching in both directions he saw green grass, trees and, further out, small buildings. Above his head he could make out small clouds. And, above that, he could see the greenery of a park on the opposite side of the giant cylinder that constituted New Washington.

"This really is beautiful, Sam. I knew what the cities were supposed to look like, but I never imagined that they would look this good."

"I know, dear. Everybody reacts that way. That's why we

bring all new colonists here first. Now, follow me." They walked on through the grass and under a few trees. Jason could hear running water in front of him and soon enough he saw two people, one large and one small, standing on the banks of a stream, fishing. It took Jason a minute to realize that he was looking at his son and his father-in-law. He and Sam stood quietly under a tree, arm in arm watching for some time before Jason moved forward to embrace his son.